RECENT ADVANCES
DIFFERENTIAL GEOMETRY

Ram Bilas Misra

CWP
Central West Publishing

RECENT ADVANCES IN DIFFERENTIAL GEOMETRY

by

Prof. Dr. Ram Bilas Misra

Ex Vice Chancellor, Avadh University, Faizabad, U.P. (India);

***Professor* of Mathematics, Research & Strategic Studies Centre,**

Lebanese French University, Erbil, Kurdistan (Iraq).

Former: *Dean*, Faculty of Science, A.P. Singh University, Rewa, M.P. (**India**);
Prof., Dept. of Maths., Higher College of Edn., Aden Univ., Aden (**Yemen**);
Professor & Head, Dept. of Maths. & Stats., A.P.S. University, Rewa, M.P. (**India**);
Prof., Dept. of Maths., College of Science, Salahaddin University, Erbil (**Iraq**);
UGC Visiting Prof., Mahatma Gandhi Kashi *Vidyapith*,Varanasi, U.P. (**India**);
Professor, Dept. of Maths, Ahmadu Bello Univ., Zaria (**Nigeria**) – designate;
Prof. & Head, Dept. of Maths. & Comp. Sci., Univ. of Asmara, Asmara (**Eritrea**);
Director, Unique Inst. of Business & Technol., Modi Nagar, Ghaziabad, U.P. (**India**);
Prof. & Head, Dept. of Maths., Phys. & Stats., Univ. of Guyana, Georgetown (**Guyana**);
Prof. & Head, Dept. of Maths., Eritrea Inst. of Technology, Mai Nefhi (**Eritrea**);
Prof.& Head, Dept. of Maths., School of Engg., Amity Univ., Lucknow, U.P. (**India**);
Prof. & Head, Dept. of Maths. & Comp. Sci., PNG Univ. of Technology, Lae (**PNG**);
Prof. of Maths., Teerthankar Mahaveer University, Moradabad, U.P. (**India**);
Prof., Dept. of Maths, Oduduwa Univ., Ipetumodu, Osun State (**Nigeria**) – designate;
Prof., Dept. of Maths, Adama Science & Technology Univ., Adama (**Ethiopia**);
Prof. & Head, Dept. of Maths. & C.S., Bougainville Inst. of Bus. & Tech., Buka (**PNG**);
Prof. & Head, Dept. of Maths., J.J.T. University, Jhunjhunu, Rajasthan (**India**);
Dean, Faculty of Science, J.J.T. University, Jhunjhunu, Rajasthan (**India**);
Professor, Dept. of Maths, Wollo University, Dessie, Wollo (**Ethiopia**);
Professor, Dept. of Appld. Maths., State Univ. of New York, Incheon (**S. Korea**);
Prof., Dept. of Maths. & Computing Sci., Divine Word Univ., Madang (**PNG**);
Director, Maths., School of Sci. & Engg., Univ. of Kurdistan Hewler, Erbil (**Iraq**);
DAAD Fellow, University of Bonn, Bonn (**Germany**);
Visiting Professor, University of Turin, Turin (**Italy**);
Visiting Professor, University of Trieste, Trieste (**Italy**);
Visiting Professor, University of Padua, Padua (**Italy**);
Visiting Professor, International Centre for Theoretical Physics, Trieste (**Italy**);
Visiting Professor, University of Wroclaw, Wroclaw (**Poland**);
Visiting Professor, University of Sopron, Sopron (**Hungary**);
Reader, Dept. of Maths. & Stats., South Gujarat University, Surat, Gujarat (**India**);
Reader, Dept. of Maths. & Stats., University of Allahabad, Allahabad, U.P. (**India**);
Asst. Prof., Dept. of Maths., College of Sci., Mosul Univ., Mosul (**Iraq**) – designate;
Senior most *NCC Officer* (Naval Wing), Univ. of Allahabad, Allahabad, U.P. (**India**);
Lecturer, Dept. of Maths., KKV Degree College, Lucknow, U.P. (**India**).

2020

This edition has been published by Central West Publishing, Australia

For more information about the books published by Central West Publishing, please visit https://centralwestpublishing.com

NATIONAL LIBRARY OF AUSTRALIA

A catalogue record for this book is available from the National Library of Australia

ISBN (print): 978-1-925823-80-6

DEDICATED TO
MY HIGHER GEOMETRY TEACHERS AND MENTORS

Dr. Hari Shankar Shukla
(Dept. of Mathematics, Kanya Kubja Degree College, Lucknow);

Prof. Dr. Murli Dhar Upadhyay – *Differential Geometry Teacher (Lucknow University, Lucknow & VC, Kumaun University, Nainital);*

Prof. Dr. R.S. Mishra – *the Ph.D. Supervisor (Head, Dept. of Mathematics, University of Allahabad, Allahabad & VC, CSM Kanpur University, Kanpur & Lucknow University, Lucknow);*

Prof. Dr. Ram Behari – *the Ph.D. thesis examiner (Head, Department of Mathematics, Delhi University, Delhi & VC, Jodhpur University, Jodhpur);*

Prof. Dr. Akitsugu Kawaguchi – *the Ph.D. thesis examiner (Nihon University, Tokyo & Hokkaido University, Sapporo, Japan);*

Prof. Dr. Wilhelm Klingenberg
(Rector, Rheinische Friedrich-Wilhelm University, Bonn, Germany – having accepted me as a DAAD-Fellow in 1972-73);

Prof. Dr. Franco Fava – *my senior co-worker & host (Department of Mathematics, University of Turin, Turin, Italy);*

Prof. Dr. Arthur Moór
(Head, Dept. of Maths., Sopron University of Forestry & Wood Sciences, Sopron (Hungary) – for his invitation in 1976);

Prof. Dr. Witold Roter
(Director, Institute of Mathematics, University of Wroclaw, Wroclaw (Poland) – for his invitations in 1980 and 1996);

Prof. Dr. Aldo Bressan
(Professor Emeritus of Rational Mechanics, Univ. of Padua, Padua (Italy) – for his invitations in 1981 and 1992);

Prof. Dr. H.D. Doebner
(Director, Arnold Sommerfeld Inst. for Mathematical Physics, Technical Univ. of Clausthal, Germany – for his invitation in 1996).

In Memoriam (Mother & Sister)

Allahabad (India) / 1964: Author with Rakesh (eldest son), mother and sister (Sarojini)

Semrai / 1973: Mrs. Sarojini Mishra
(born 31.7.1948 - died 1.8.1980)

CONTENTS

PREFACE

The present book is author's life-time's vision. It finds its base in the research oriented academicians' pursuits in diverse fields. The dedication of India's celebrated Hindi poet (Saint Tulsidas), French geometer Professor Elié Cartan and the author's parents in their respective domains ever impressed and inspired him to accomplish this long due project. It was started as early as in 1976 during my stay at University of Turin, Turin (Italy), where a chapter on Finsler Geometry offering a basic course was completed. It remained so for a decade and, later, it was brought out in the Internal Reports of the ICTP, Trieste (Italy) in 1986. The same with some improvements is included here as Chapter 4. Perhaps, it is the divine will to get the book accomplished in the birth centenary year of my research supervisor (*Padmashree* Prof. Dr. R.S. Mishra) only.

It narrates the saga of origin of geometry especially non-Euclidean and covers up to few generalizations of non-Euclidean geometry such as Riemannian, Finslerian, Minkowskian, Kawaguchi geometry, Geometry of Complex spaces, etc. However, more attention is focused on the recent developments in vivid models of Finslerian structures. The book provides brief account of a large number of contributions made in the field by authors at large and it may introduce the subject to the learners and researchers in the field.

The book has two diverse components: Chapters 1 - 3 and 14 - 15 are mainly expository without rigorous treatment of mathematical formulation in order to stimulate the new readers. In between, the Chapters 4 - 13 deal with the Finslerian Geometry rigorously. Altogether, it contains *fifteen* chapters of which the first one summarizes the developments of non-Euclidean geometry and its generalizations. The saga of "Differential geometry: its past and future" is presented in the second chapter while the next one discusses metrics of curved surfaces and various spaces. Non-Euclidean, Riemannian and its generalizations, Finslerian, generalized (Kawaguchi, Areal and Kähler) as well as special Finslerian (Berwald, Landsberg, Randers', Kropina and other spaces introduced by Matsumoto [73][1)], [83]) spaces are covered.

Chapter 4 offers a detailed course on Finsler Geometry and can be read as a sequel to the books by Rund [189] and Yano [234]. Starting *ab initio* metric, theory of connections (both depending upon the metric and non-metric approaches of Rund and Berwald) giving rise to different covariant derivation processes and commutation rules, theory of curvatures: both involving Cartan's approach and that of Berwald have been included. The projective transformation and the corresponding projective curvature tensors, conformal transformations and their applications to certain types of Special Finslerian models are presented in the Chapter 5. Theory of Lie derivation and corresponding groups of transformations: motions, affine motions, projective motions, conformal motions and curvature collineations are introduced in Chapter 6. Chapter 7 deals with the theory of Symmetric and Projectively symmetric Finsler spaces; while Spaces of recurrent curvature are covered in Chapter 8. Various groups of transformations are also included in these chapters. However, projective motions in recurrent spaces are dealt separately in Chapter 9. Next chapter offers a brief account of a large number of explorations in the theory of Groups of Transformations in various models of Finslerian structures notably bi-symmetric and bi-recurrent Finsler spaces. The chapter concludes with the information of different types of transformations studied in various Finslerian models in a tabular form. Spaces with zero projective curvature, so-called projectively flat spaces are studied in Chapter 11. The next two chapters deal with the concircular infinitesimal transformations, introduced by Yano [233], in Finslerian spaces. These three chapters (11 – 13) detail the author's own researches.

Chapter 14 offers an account of axiomatic approach to tensors. Vector spaces (spanned by contravariant vectors) and its dual (vector) space equipped with covariant vectors and their tensor product spaces are considered. It ends with different types of tensors. The last chapter reports about various "Physical Field Theories" touching upon Young-Mills SU(2) and SU(3) symmetries. Certain applications of Finslerian models are suggested: for instance, Finslerian Physics, Mechanics of open systems and Thermodynamic Finsler spaces. The works of Antonelli, Ingarden and Matsumoto [7], Antonelli and R. Miron [9], G.S. Asanov [13], A. Bejancu [17], Ingarden and Nakagomi [51], Mrugala [154], etc. are recommended for further reading.

The contents are divided into Sections and the discussion within the Sections is presented in the form of Definitions, Theorems, Corollaries, Notes and Examples. The sub-titles within the Sections are numbered in decimal pattern. For instance, the equation number ($c.s.e$) refers to the eth equation in the sth section of Chapter c. When c coincides with the chapter at hand, it is dropped. Foot-notes are numbered serially irrespective of concerned chapter. Adequate references to the results appeared earlier are made in the text avoiding their unnecessary repetition. Double slashes marked at the end of Theorems, Corollaries, etc. indicate their completion. For brevity, some set-theoretic notations and symbols are frequently used, e.g. the symbol $\Rightarrow$ means *implies*. The *logarithm* of a number to the exponential base e is denoted by ln. Latin mathematical symbols, if indicating a number, set or space are *italicized*; but are kept in normal (non-italic) fonts for objects such as point, curve, surface, etc. Vectors are always denoted by Latin symbols in normal fonts but in bold face type. Their Greek counterparts are usually in normal fonts. A comprehensive bibliography of the subject is provided. Symbols and notations used in the book are enlisted with adequate references to the text. An alphabetical index is also supplemented at the end making access to the contents easier.

Author's long teaching career of more than ***five decades*** at various universities round the globe and research expertise in different fields helped him for lucid presentation of the subject. It is dedicated to his teachers of higher geometry and mentors in global interaction. Sincere gratitude is also offered to various Universities all over the world especially University of Allahabad, Prayagraj (India); University of Bonn, Bonn (Germany); University of Turin, Turin (Italy); Abdus Salam International Centre for Theoretical Physics, Trieste (Italy); University of Guyana, Georgetown (Guyana); P.N.G. University of Technology, Lae (PNG); Adama Science & Technology University, Adama (Ethiopia); State University of New York, Incheon (South Korea); Divine Word University, Madang (PNG); University of Kurdistan Hewler, Erbil and Lebanese French University, Erbil (Iraq). The family members also deserve sincere thanks for their consistent cooperation providing enough time for my academic pursuits. Sincere thanks are also due to the publisher for their valuable cooperation for bringing the book into limelight in a limited time.

Although proofs are read with utmost care yet an oversight or any discrepancy brought to the notice of the author by the inquisitive readers(s) shall be thankfully acknowledged. What a coincidence of finalizing the first draft of the book on the day (1st August) when author's only surviving (younger) sister Sarojini breathed her last in 1980. Giving a second thought, Chapters 6 - 9 were planned and completed later.

Lucknow (India): September 29, 2019 Ram Bilas Misra

[1)] Numbers within square brackets refer to the references given at the end.

CHAPTER 1

NON-EUCLIDEAN GEOMETRY AND ITS GENERALIZATIONS

§ 1. The origin of geometry

The word "Geometry" is derived from the Greek word "*Geometria*" which means "to measure". The mensuration of land and the fixing of boundaries necessitated by the repeated inundations of the Nile gave birth to geometry. The credit of introducing the study of geometry goes to Thales of Miletus (640-546 B.C), one of the seven "wise men" of Greece. This marks the first stage in the evolution of geometry. Pythagoras (580-500 B.C) interpreted geometry as a metrical science. About 300 B.C, attempts by Euclid to present the geometrical propositions as based upon a few axioms and definitions became so famous that he became as popular as Aristotle himself. Euclid gave five postulates. The first three of them refer to the construction of straight lines and circles. The fourth asserts the equality of all right angles and the fifth is the famous parallel line postulate:

"*If a straight line falling on two straight lines make the interior angles on the same side less than two right angles the straight lines, if produced indefinitely, meet on that side on which are the angles less than two right angles.*"

Savile [194], D'Alembert [34] and many others attempted to prove the above postulate but without success. Ludlam [69] gave the axiom equivalent to that of Euclid's above postulate:

"*Two intersecting lines cannot both be parallel to the same straight line*", or equivalently "*through a given point not more than one parallel line can be drawn to a given straight line*".

The geometry based on Euclid's axioms is called *Euclidean geometry*.

§ 2. First glimpse of Non-Euclidean geometry

Saccheri (1677- 1733) devised an entirely different mode of attacking the problem in an attempt to institute a *redactio ad absurdum* [191].

In those days the favorite starting point was the conception of parallels as equidistant straight lines. But Saccheri, like some of his predecessors, observed that it would not do to assume it in the definition. He proposed the Hypothesis of the:

(i) Right Angle, (ii) Obtuse Angle, (iii) Acute Angle

and established a number of theorems, of which the following are important:

Theorem 2.1. *If one of the three hypotheses is true in one case, the same hypothesis is true in any case.*

Theorem 2.2 *On the hypothesis of the* $\left\{\begin{array}{l}\textit{right}\\ \textit{obtuse}\\ \textit{acute}\end{array}\right.$ *angle the sum of the angles of the triangle is* $\left\{\begin{array}{l}\textit{equal to}\\ \textit{greater than}\\ \textit{less than}\end{array}\right.$ *two right angles.*

Afterwards Saccheri demolished the hypothesis of the obtuse angle by showing that it contradicts Euclid's result: "*The sum of any two angles of a triangle is less than two right angles*". He requires nearly twenty more theorems before he can demolish the hypothesis of the acute angle. In spite of all his efforts, however, he does not seem to be quite satisfied with the validity of his proof and he offered another proof in which he lost himself, like many others. If Saccheri had had a little more imagination and been less bound down by tradition, he would have anticipated by a century the discovery of the two Non-Euclidean geometries which follow from his last two hypotheses.

Gauss [2)] (1777 - 1855) [44] was probably the first to obtain a clear idea of a geometry other than that of Euclid. The new geometry was called, by Gauss, *an Anti-Euclidean*, and finally, *Non-Euclidean*. In 1818 Gauss received a letter from Schweikart (1780 - 1859) explaining a system of geometry which latter called *Astral Geometry*. In this geometry *the sum of the angles of a triangle is always less than two right angles*. Schweikart did not publish any account of his researches, but he induced his nephew, Taurinus (1794 - 1874) [223], to take up the question. Though the ideas of his uncle could not appeal to Taurinus but a few years' later he attempted a treatment of the theory of parallels

[223]. He worked out some of the most important trigonometrical formulae for Non-Euclidean geometry by using the fundamental formulae of spherical geometry with an imaginary radius. He expressed his results in terms of logarithms and exponentials and called his geometry the *Logarithmic Spherical Geometry.* Though Taurinus may be regarded as an independent discoverer of Non-Euclidean trigonometry, he always believed, unlike Gauss and Schweikart, that Euclidean geometry was necessarily the true one. In the last days of his life when Taurinus found that his work attracted no attention he became so disappointed that he burnt the remainder of the edition of his *Elementa,* which is now one of the rarest of books.

F.L. Wachter (1792 - 1817), a student of Gauss, also denied Euclid's postulate. It is remarkable that he affirms, that even if the postulate be denied, the geometry on sphere becomes identical with the Euclidean geometry when the radius is indefinitely increased, though it is distinctly shown that the limiting surface is not a plane. This is one of the greatest discoveries of Lobachevski and Bolyai. If Wachter could survive a little more he might have been the discoverer of Non-Euclidean geometry, for his insight into the question was far beyond that of an ordinary parallel postulate demonstrator.

§ 3. The discovery of Non-Euclidean geometry

In 1826 Nikolai Ivanovich Lobachevsky (1793 - 1856) read a paper entitled "*Exposition succinate des principles de la gèomètrie avec une dèmonstration rigoureuse du thèorem des parallèles*" in the Physical and Mathematical Section of the University of Kazan. In this paper he explained the principles of his *Imaginary geometry* which is more general than that of Euclid. In this new geometry:

(i) *Two parallels can be drawn to a given line through a given point,* and

(ii) *The sum of the angles of a triangle is always less than two right angles.*

Lobachevsky wrote some half dozen extensive memoirs expounding the new geometry. The first of these was in Russian. In 1840, he wrote a small book entitled "*Geometrische Untersuchungen zur Theorie der Parallelinien*" and just before his death he wrote a summary of his researches under the title *Pangeometry* in French.

In 1804 Wolfgang Bolyai (1775 - 1856), a fellow-student and friend of Gauss at Göttingen, sent to the latter a note on "Theory of Parallels". In this note he tried to prove that a series of equal segments placed end to end at equal angles like the sides of a regular polygon, must make a complete circuit. Meanwhile, his son John Bolyai (1802-1860) attempted to find a proof for the parallel postulate and gradually the results which would follow from the denial of the axiom developed in his mind the idea of a general or *Absolute* geometry. This geometry would contain the ordinary or Euclidean geometry as a special case. In 1823, John Bolyai had worked out the main ideas of the Non-Euclidean geometry. Through his letter of 3rd November, 1823 he announced to his father:

"I have made such wonderful discoveries that I am myself lost in astonishment and it would be an irresponsible loss if they remain unknown. Dear father ! when you read them, you too will acknowledge it. I cannot say more now except that out of nothing I have created a new and another world".

John Bolyai's presentation was truer than he suspected, for at this very moment Lobachevsky at Kazan, Gauss at Göttingen, Taurinus at Cologne were all on the verge of this great discovery. In spite of the correctness John Bolyai's work was published after a long gap of nine years in Tentamen under the title "Appendix, Scientiam spatii absolute veram exhibens".

John Bolyai, Lobachevsky and Gauss arrived at their ideas independently. But as Bolyai and Lobachevsky possessed the convictions and the courage to publish their results, which Gauss lacked, to them alone has gone the credit of the discovery. The ideas given by Lobachevsky and John Bolyai attained wide recognition only after Baltzer gave attention to them in 1867. It is remarkable that while Saccheri and Lambert both considered the two hypotheses, it never occurred to Lobachevsky, Bolyai, Gauss, Schweikart, Taurinus and Wachter to admit the hypothesis that the sum of the angles of a triangle may be greater than two right angles. This involves the conception of a straight line as being unbounded but yet of finite length. It was only B. Riemann (1826 - 1866) who in his famous lecture of 1854 [185], delivered at Göttingen, pointed out the distinction between unboundedness and infiniteness.

Klein [58] called the geometry of $\begin{cases}\text{Lobachevsky}\\ \text{Riemann}\\ \text{Euclid}\end{cases}$ as $\begin{cases}\text{hyperbolic}\\ \text{elliptic}\\ \text{parabolic}\end{cases}$.

§ 4. Geometry on different surfaces

Through two points A and B on a surface there will generally pass one definite line belonging to the surface, namely, the shortest distance between the two points. This line is called the *geodesic* joining the two points. On a sphere the geodesic joining the two points is an arc of the great circle through them.

If, however, we desire to compare the geometry on a surface with the geometry on a plane, it is natural to make the geodesics, which measure the distances on one surface, correspond to the straight lines of the other. It is also natural to call the two figures traced upon the surface as geodetically equal when there is a point to point correspondence between them such that the geodesic distances between corresponding points are equal. If the surface is made of a flexible and inextensible sheet, we obtain a representation of this conception of equality. Then by the movement of the surface, which does not remain rigid, but is bent as described above equal figures are to be superposed one upon the other.

Let us consider a piece of cylindrical surface. By simple bending, without stretching, folding or tearing this can be applied to a plane area. In this case two figures ought to be called equal on the surface, which coincide with equal areas on the plane though, of course, two such figures are not in general equal in space. Thus upon any cylindrical surface we will have a geometry similar to that on any plane, and in general, upon any developable surface.

But the geometry on the sphere is essentially different from that on the plane, since it is impossible to apply a portion of the sphere to the plane. However, there is an analogy between the geometry on the plane and the geometry on the sphere. This analogy is that the sphere can be freely moved upon itself, so that propositions in every way analogous to the postulates of congruence on the plane hold for equal figures on the sphere.

In order that a suitable bounded surface, by bending but without stretching, can be moved upon itself in the same way as a plane, certain number K must have a constant value at all points of the surface. This number was introduced by Gauss and is called after him as *Gaussian curvature*.

The following table will establish the analogy between the geometry on a surface of constant Gaussian curvature and that on a portion of a plane:

Geometry on the surface of constant Gaussian curvature	Geometry on a portion of a plane
1. Point	1. Point
2. Geodesic	2. Straight line
3. Arc of geodesic	3. Rectilinear segment
4. Fundamental properties of the equality of geodesic arcs and angles	4. Postulates of congruence for rectilinear segments and angles
5. Sum of the angles of geodesic triangle *ABC* is given by $A + B + C = \pi + \int_{ABC} K\, dS,$ where the surface integral has been taken over the geodesic triangle *ABC*.	5. Sum of the angles of a rectilinear triangle is equal to two right angles

There arise the following three cases for the sum of the angles of a triangle in the geometry on a surface of constant curvature:

Case 1. When the surface is synclastic so that its Gaussian curvature is positive, say $K = 1 / k^2 > 0$, we have

$$\int_{ABC} K\, dS = (1/k^2) \int_{ABC} dS.$$

But $\int_{ABC} dS$ = area of the triangle ABC, say Δ. Therefore

$$\angle A + \angle B + \angle C = \pi + \Delta / k^2. \tag{4.1}$$

Thus, *the sum of the angles of the geodesic triangle on such a surface is greater than two right angles.*

Case 2. When the surface is anticlastic so that its Gaussian curvature is negative, say $K = -1 / k^2$, we have

$$\angle A + \angle B + \angle C = \pi - \Delta / k^2.$$

Hence, *the sum of the angles of the geodesic triangle on such a surface is less than two right angles.*

Case 3. When surface is developable whose Gaussian curvature is zero the *sum of angles of a triangle drawn on that will be equal to two right angles.* For example, a cylinder and cone are developable surfaces Minding (1806 - 1885) was the first to begin the study of the surfaces of constant negative curvature with the investigation of the surfaces of revolution to which they could be applied. The following remark of Minding, fully proved by D. Codazzi (1824 - 1873), establishes the trigonometry of such surfaces.

Minding's remark. In the formulae of spherical trigonometry let the angles be kept fixed and the sides multiplied by $i = \sqrt{(-1)}$. Then we obtain the equations which are satisfied by the elements of the geodesic triangle on surfaces of constant negative curvature.

Note. Such a trigonometry was called the *pseudo-spherical trigonometry.*

In this way the geometry upon a surface of constant positive or negative curvature can be considered as a concrete interpretation of the Non-Euclidean geometry obtained in a bounded plane area. The possibility of interpreting the geometry of a two-dimensional manifold by means of ordinary surfaces was observed by Riemann.

§ 5. The generalizations of Non-Euclidean geometry

In the famous memoir of 1854 Riemann discusses various possibilities by means of which an n-dimensional manifold may be endowed with a metric. He paid particular attention to a metric defined by the positive square root of a positive definite quadratic differential form. Thus, the foundations of Riemannian geometry were laid, nevertheless, it is also suggested that the positive fourth root of a fourth order differential form might serve as a metric function. Thus, the distance ds between the neighboring points x^i and $x^i + dx^i$ is represented by

$$ds = F(x^i + dx^i), \qquad (i = 1, 2, \ldots, n),$$

where the metric function $F(x^i + dx^i)$ is positive, homogeneous of first degree in differentials dx^i and convex. The first systematic study of manifolds endowed with such a metric was delayed by more than six decades. It was an investigation of this kind which formed the subject matter of the thesis of P. Finsler in 1918. Later on such spaces were eventually named after him as Finsler spaces. Since then work has extensively been done in Finsler geometry. A large number of works in this direction has been referred to in the end. The Finsler geometry has been further generalized. The following are the main generalizations:

1. Cartan Spaces,
2. Kawaguchi spaces,
3. The generalized variations spaces,
4. Landsberg spaces,
5. Infinitely - dimensional Finsler spaces,
6. Areal Spaces.

Foot-note:

2) Gauss published nothing on the subject except a few reviews. It is clear from his correspondence and notes that he was deeply interested in it. He refrained from publishing anything because he feared the clamour of Beotians. Indeed, at that time the problem of parallel lines was greatly discredited and anyone who occupied himself with it was liable to be considered as a crack.

CHAPTER 2

DIFFERENTIAL GEOMETRY: ITS PAST AND FUTURE

§ 1. Introduction

Differential Geometry has a long history and has been widely explored for the past more than two centuries. Recent advances in the fields of topology and abstract algebra which, by now, have undoubtedly established their dominance over almost all disciplines in pure mathematical sciences gave tremendous impetus to *differential geometry*. The present day differential geometry is far from the cries of Ricci's tensor analysis initiated in the beginning of the 20th century; and can now be well regarded as *differential topology*. Thus, in the realms of contemporary analysis the development of the subject is relatively new and is most suitable field of pursuit. This development, however, has not been as abrupt as might be imagined from a reading of the subject. It has its roots in the movement towards *differential geometry in the large* to which geometers such as E. Hopf and W. Rinow, M. Cohn-Vossen, G. de Rham, W.V.D. Hodge, and S.B. Myers gave impetus. The objective of their work was to derive relationships between the topology of a manifold and its local differential geometry. Other sources of inspiration were Elié Cartan (whose fundamental contributions to *exterior differentiation* could be recognized by many only after his death) and M. Morse's *calculus of variations in the large*. One of the major new ideas was that of a *fibre bundle* which gave a global structure to a differentiable manifold more general than that included in the older theories. Methods of differential geometry were applied with outstanding success to the theories of complex manifolds and algebraic varieties and those in turn have stimulated *differential geometry*. The discovery of invariants of the differential structure of a manifold, which are not topological invariants, by J. Milnor established *differential topology* as a discipline of major importance.

§ 2. Emergence of generalized metric structures

A geometry different than that of Euclid, namely *Non-Euclidean geometry*, was introduced by celebrated mathematician: Carl Friedrich Gauss in the 19th century. A special case of non-Euclidean geometry endowed with a positive definite real symmetric metric was discussed by Bernhard Riemann in his famous *habilitation* lecture "*Über die Hypothesen, welsche der Geometrie zugrunde liegen*" delivered at Göttingen in 1854. This led to the discovery of a new geometry: *Riemannian geometry* - called after him. Riemann also discussed various other possibilities by means of which an n-dimensional space may be endowed with a metric. Amongst these there was also a case in which the positive fourth-root of a fourth degree differential form defines a metric. However, a systematic study of manifolds endowed with such a metric was delayed by more than six decades. It was an investigation of this kind in 1918 which formed the subject matter of Paul Finsler's Ph.D. dissertation submitted to the University of Basel (Switzerland) after whom such manifolds were eventually named. This created a new horizon in the name of *Finsler geometry* in the field of differential geometry for its exploration. Soon after publication of Finsler's thesis the work in this direction was taken up by several geometers: Elié Cartan, L. Berwald, Enrico Bompiani, J.A. Schouten, J. Douglas, W. Barthel, Hanno Rund, S. Golab, V.V. Vagner, Otto Varga, Andre Lichnerowicz, Akitsugu Kawaguchi, E.T. Davies, H. Busemann, Arthur Moór, K. Takano, Kentaro Yano, S.S. Chern, M.S. Knebelman and others. They have been the pioneers in the field.

§ 3. Finslerian structures

Investigations in the geometry of Finsler manifolds are not merely generalizations of corresponding results of Riemannian geometry; rather are, at times, very much different to their counterparts in Riemannian geometry. According to Rund, it is a branch of differential geometry that is closely related to various topics in theoretical physics, notably analytical dynamics and geometrical optics. Unlike the Riemannian case having only one symmetric *linear connection* and one (corresponding) curvature tensor, there exist various types of linear connections and the corresponding curvature tensors in Finsler geometry. Also, in the theory of sub-manifolds of a Finsler manifold there exist two distinct types of sets of unit normal vectors and two different connections: *induced* and *intrinsic* giving rise to different processes of covariant differentiation while in a Riemannian sub-manifold there is only one set of unit normals and only one type of connection parameters.

Various types of transformations: (i) infinitesimal point transformations, (ii) conformal transformations, and (iii) projective transformations of the Riemannian geometry were extended to Finsler geometry in similar ways yet the theory of Lie differentiation in the two geometries differs to a great extent. Besides, in Finsler manifolds, infinitesimal transformations have also been studied in a general form considering the generating vector of the transformation as a function of the line-element instead of being merely a point function. This work was done, for the first time, by Buchin Su [218], the present author with R.S. Mishra [95], and the author [105], [107].

§ 4. Symmetric and recurrent Finsler manifolds

The concept of *Kappa spaces* introduced by H.S. Ruse [190], and later extended by A.G. Walker [230] in the name of *Recurrent Riemannian spaces* has been further extended to Finsler geometry and to the generalized metric and non-metric spaces. A class of Finsler manifolds, in which the covariant derivative of a non-null curvature tensor is expressible as a product of the curvature tensor and an arbitrary non-null covariant vector field, is said to be *recurrent*. Such manifolds are studied by A. Moór [147], R.S. Mishra and H.D. Pande [97], R.N. Sen [196], the present author [110], [112], [113], the author with F.M. Meher [128], [130], [133] and others.

Another class of Finsler manifolds wherein the curvature tensor or projective curvature tensor possess vanishing covariant derivatives is studied by the author [108], [109]. Such manifolds are called *symmetric* or *projectively symmetric Finsler manifolds* respectively as per the terminology used by Elié Cartan [31] for a similar class of Riemannian manifolds.

Infinitesimal transformations preserving parallelism of a pair of vectors are called *affine motions* and have been studied for above kind of Finsler manifolds by A. Kumar [64], the present author [111], the author with F.M. Meher [128], [130], [134] and the author with R.S. Mishra and Nawal-Kishore [96].

Infinitesimal transformations preserving the geodesic character of curves define *projective motions*. Such transformations are studied, in different types of Finsler manifolds discussed above, by the present author [112], [113], the author with F.M. Meher [127], [131], [132], the author et al. [141], H.D. Pande with J.P. Pandey [169], F.M. Meher [88], P.N. Pandey [170], [180], O.P. Singh [203], and R.S. Sinha [214], [215].

Study of motions, affine motions, projective motions, conformal and homothetic motions generated by infinitesimal transformations of different types in Finsler manifolds is very vast and fruitful for its exploration.

§ 5. Techniques and diverse approaches in differential geometry

There have been three techniques of computation in use in differential geometry, namely: (i) classical tensor approach (with indices), (ii) exterior differential calculus approach of Elié Cartan, and (iii) formalism of covariant differentiation $\nabla_X Y$. The last one is the latest amongst these. Following abstract methods and employing the newest technique much has been done in the geometry of Riemannian manifolds, mainly in the European and American schools, but the study of Finsler manifolds in context with this logical approach has been done by Japanese geometers: Makoto Matsumoto [73], [76], [83]; K. Yano [235]; K. Yano and Y. Muto [237]; K. Yano and T. Okubo [238]; Hassan Akbar-Zadeh and Anna Wegrzynowska [3] and others.

CHAPTER 3

METRICS OF CURVED SURFACES AND SPACES

§ 1. Introduction

The word *metric* is derived from the Greek word *metria* meaning *measurement*. Historically, its origin lies in the measurement of distances on the surface of the Earth so that inundations in the river Nile could be curbed out.

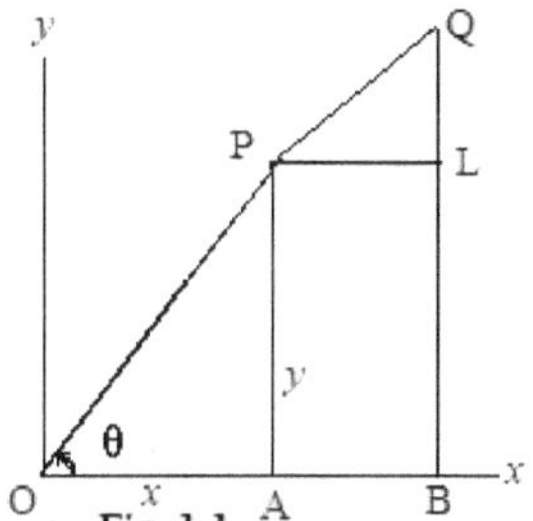

Fig. 1.1

Before introducing metrics of a curved surface and spaces, we examine the formula for linear distance between two points in a plane, which is essentially based on the celebrated Pythagoras theorem. Given two infinitesimal points P, Q with rectangular Cartesian coordinates (x, y) and $(x + dx, y + dy)$ in a plane the distance PQ $\equiv ds$ between them is measured by

$$(ds)^2 = (\mathrm{PL})^2 + (\mathrm{LQ})^2 = (\mathrm{OB} - \mathrm{OA})^2 + (\mathrm{BQ} - \mathrm{BL})^2 = (dx)^2 + (dy)^2. \quad (1.1)$$

The right member of above equation is called the *metric* of the Euclidean plane E_2. Above formula can be also generalized to a three-dimensional Euclidean space E_3 as well as to a Euclidean space E_n of arbitrary dimension $n \varepsilon \mathrm{N}$.

In the next Section, we first obtain a formula for the metric of a curved surface V_2 immersed in the space E_3 and then deduce Eq. (1.1) from the same as a special case.

§ 2. Metric of a V_2

Let V_2 be a curved surface immersed in the Euclidean space E_3. The position vector **r** of a point P ε V_2 with respect to the origin O in E_3, is thus, expressible as a function of two independent parameters, say u^1 and u^2:

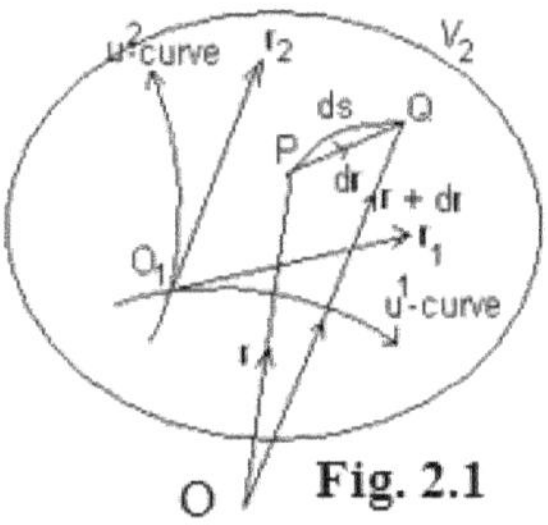

Fig. 2.1

$$\mathbf{r} = \mathbf{r}\,(u^1, u^2). \quad (2.1)$$

The sets of points of V_2 for which one of the parameters remains constant describe the parametric curves:

u^1-curve (for which u^2 = const.), u^2-curve (for which u^1= const.).

Thus, the parametric curves having equations

$$u^2 = \text{const.} \qquad \text{and} \qquad u^1 = \text{const.} \tag{2.2}$$

respectively form a curvilinear coordinate system on V_2. Accordingly, Eq. (2.1) determines the position vectors $\mathbf{r} = \mathbf{r}\,(u^1)$ and $\mathbf{r} = \mathbf{r}\,(u^2)$ of the points on respective parametric curves. The derivative of $\mathbf{r}$ with respect to a parameter defines the tangent vector to the concerned curve. Thus, the vectors

$$\mathbf{r}_{,1} = (\partial \mathbf{r} / \partial u^1)_{u2\ = \text{const.}} \quad \text{and} \quad \mathbf{r}_{,2} = (\partial \mathbf{r} / \partial u^2)_{u1\ = \text{const.}} \tag{2.3}$$

are tangential to the respective parametric curves at the point O_1.

If Q (with the position vector $\mathbf{r} + d\mathbf{r}$) is a neighbouring point of P on the surface, the infinitesimal vector $\overrightarrow{PQ} \equiv \overrightarrow{OQ} - \overrightarrow{OP} \equiv d\mathbf{r}$ is, thus, obtained from Eq. (2.1):

$$d\mathbf{r} = \mathbf{r}_{,1}\, du^1 + \mathbf{r}_{,2}\, du^2; \tag{2.4a}$$

where the Taylor's theorem of Differential Calculus has been applied for the vector function $\mathbf{r}$ to the first order of approximation. Employing the *Einstein's summation convention*: "A dummy index is ought to be summed over its specified range even without writing the summation symbol $\sum$" Eq. (2.4a) may be more compactly written as

$$d\mathbf{r} = \mathbf{r}_{,\lambda}\, du^\lambda, \qquad \lambda = 1, 2. \tag{2.4b}$$

Forming the scalar product of Eq. (2.4b) with itself and using the fact that in limiting case, the length of the chord PQ equals the arc-length PQ, we obtain

$$(ds)^2 = d\mathbf{r}.\, d\mathbf{r} = (\mathbf{r}_{,\lambda}\, du^\lambda) \,.\, (\mathbf{r}_{,\mu}\, du^\mu) = g_{\lambda\mu}\, du^\lambda du^\mu, \tag{2.5}$$

where

$$g_{\lambda\mu} \equiv \mathbf{r}_{,\lambda} \cdot \mathbf{r}_{,\mu} \tag{2.6}$$

are functions of the parameters u^1, u^2; and are called the *fundamental magnitudes of first order* for V_2. Thus, the right member of Eq. (2.5) provides a natural metric on the surface V_2. The scalar product being commutative the functions $g_{\lambda\mu}$ are symmetric in their indices λ, μ. Hence, the metric of V_2 is symmetric. Further,

$$g_{11} \equiv \mathbf{r}_{,1} \cdot \mathbf{r}_{,1} = |\,\mathbf{r}_{,1}\,|^2, \quad \text{and} \quad g_{22} \equiv \mathbf{r}_{,2} \cdot \mathbf{r}_{,2} = |\,\mathbf{r}_{,2}\,|^2 \tag{2.7}$$

being positive determine the magnitudes of the vectors $\mathbf{r}_{,1}$ and $\mathbf{r}_{,2}$

$$\sqrt{(g_{11})} = |\,\mathbf{r}_{,1}\,| \qquad \text{and} \qquad \sqrt{(g_{22})} = |\,\mathbf{r}_{,2}\,|. \tag{2.8}$$

If the parametric curves are inclined at an angle ω, the Eq. (2.6) also determines

$$g_{12} \equiv \mathbf{r}_{,1} \cdot \mathbf{r}_{,2} = |\,\mathbf{r}_{,1}\,|\,|\,\mathbf{r}_{,2}\,| \cos\omega = \sqrt{(g_{11}\, g_{22})} \cos\omega. \tag{2.9}$$

Hence,

$$g_{11} g_{22} - (g_{12})^2 = g_{11} g_{22} (1 - \cos^2\omega) = g_{11} g_{22} \sin^2\omega \tag{2.10}$$

is positive, so it may be denoted by g^2. Thus, the discriminant

$$g^2 \equiv g_{11} g_{22} - (g_{12})^2 \tag{2.11}$$

of the metric in Eq. (2.5) is positive; and, therefore, the matrix of the metric is symmetric and non-singular.

From Eq. (2.9), it follows that $g_{12} = 0$ is a necessary and sufficient condition for orthogonality of the parametric curves. Accordingly, the metric of V_2 equipped with orthogonal parametric curves reduces to

$$(ds)^2 = g_{11}\,(du^1)^2 + g_{22}\,(du^2)^2. \tag{2.12}$$

2.1. Viewing the Euclidean plane E_2 on the lines of above discussion the position vector $\mathbf{r}$ of the point P (cf. Fig. 1.1) is expressible as

$$\mathbf{r} \equiv \overrightarrow{\mathrm{OP}} = \overrightarrow{\mathrm{OA}} + \overrightarrow{\mathrm{AP}} = x\,\hat{\mathbf{i}} + y\,\hat{\mathbf{j}}, \tag{2.13}$$

where $\hat{\mathbf{i}}$, $\hat{\mathbf{j}}$ are the unit vectors acting along the respective coordinate axes Ox and Oy. Noting the invariance of one of the variables ($y = 0$ along x-axis and $x = 0$ along y-axis) and taking the partial derivatives of Eq. (2.13) with respect to these variables:

$$\mathbf{r}_{,1} = (\partial\mathbf{r}/\partial x)_{y\,=\,\text{const.}} = \hat{\mathbf{i}}, \quad \text{and} \quad \mathbf{r}_{,2} = (\partial\mathbf{r}/\partial y)_{x\,=\,\text{const.}} = \hat{\mathbf{j}}; \tag{2.14}$$

we obtain

$$d\mathbf{r} = (\partial\mathbf{r}/\partial x)\, dx + (\partial\mathbf{r}/\partial y)\, dy = (dx)\,\hat{\mathbf{i}} + (dy)\,\hat{\mathbf{j}}. \tag{2.15}$$

Squaring this vector and noting orthogonality of the unit vectors $\hat{\mathbf{i}}$ and $\hat{\mathbf{j}}$, we derive Eq. (1.1). Thus, for E_2, we have

$$\left.\begin{aligned} g_{11} &\equiv \mathbf{r},_1 . \mathbf{r},_1 = 1, \quad g_{12} \equiv \mathbf{r},_1 . \mathbf{r},_2 = 0 = g_{21}, \\ g_{22} &\equiv \mathbf{r},_2 . \mathbf{r},_2 = 1, \quad g^2 \equiv g_{11} g_{22} - (g_{12})^2 = 1. \end{aligned}\right\} \tag{2.16}$$

Accordingly, the matrix of coefficients in an Euclidean metric is the identity matrix:

$$((g_{\lambda\mu})) = \begin{bmatrix} 1 & 0 \\ 0 & 1 \end{bmatrix}. \tag{2.17}$$

In other words, for a special choice of the metric coefficients

$$g_{\lambda\mu} = \delta_{\lambda\mu} = 1 \text{ (when } \lambda = \mu), \quad 0 \text{ (when } \lambda \neq \mu); \tag{2.18}$$

the metric in Eq. (2.5) reduces to the Euclidean metric in Eq. (1.1):

$$(ds)^2 = (du^1)^2 + (du^2)^2.$$

§ 3. Non-Euclidean metric

The natural metric in Eq. (2.5) of a V_2 may be generalized in many ways. If the functions $g_{\lambda\mu}$ are no more derivable according to Eq. (2.6) and are chosen arbitrarily as functions of the parameters they are not necessarily symmetric in λ, μ. As such, Eq. (2.5) then provides a general metric of a V_2:

$$(ds)^2 = g_{\lambda\mu}\, du^\lambda du^\mu. \tag{3.1}$$

The right member of above equation is called a *non-Euclidean metric* and the surface V_2 equipped with such a metric as *non-Euclidean surface*.

§ 4. Riemannian metric

Extending above discussion to a space V_n of arbitrary dimension $n \in \mathbb{N}$ and choosing a curvilinear coordinate system given by an n-tuple $(x^1, x^2, \ldots, x^n)$, briefly denoted by (x^i), $i = 1, 2, \ldots, n$, the formula (3.1) assumes the form

$$(ds)^2 = g_{ij}\, dx^i dx^j, \tag{4.1}$$

where the functions g_{ij} are arbitrary functions of the coordinates (x^k). However, imposing symmetry of g_{ij} in their indices i, j and assuming the right member of Eq. (4.1) as a positive definite quadratic differential form in the (real) variables x^i's, the Eq. (4.1) defines *Riemannian metric*. The space V_n equipped with such a metric is then called a *Riemannian space*. This generalization of a metric was, for the first time, given by a German geometer Bernhard Riemann (1826 -1866 A.D.) in his famous lecture: "*Über die Hypothesen, welsche der Geometrie Zugrunde liegen*" (i.e. the hypotheses on which the foundations of geometry are based) delivered at Göttingen (Germany) in the year 1854. He discussed various possibilities of a metric by which a curved space can be equipped with but he confined himself to the study of a Riemannian metric only. He also expressed how to measure the infinitesimal (arcual) distance *ds* between two infinitesimal points by a positive fourth-root of a fourth degree differential form that eventually led to a *Finslerian metric*. However, the study of such a metric was delayed by over six decades. It was a Swiss geometer, Paul Finsler, who took up the discussion of such a metric and submitted his doctoral dissertation at the University of Basel (Switzerland) in 1918.

§ 5. Generalized Riemannian metric

Similar to non-Euclidean metric: relaxing the symmetry of a Riemannian metric but retaining the positive definite character of the metric in Eq. (4.1), J.A. Schouten [195] considered a metric as in Eq. (4.1) with g_{ij} as *arbitrary functions of* x^i*'s* which need not be symmetric in i, j. He called such a metric as *generalized Riemannian metric* and the space V_n equipped with such a metric as *generalized Riemannian space*.

§ 6. Semi-symmetric Riemannian metric

Let V_n be an n-dimensional Riemannian space with curvature tensor R^i_{jkh} arising from the covariant differentiation process. If there holds the relation

$$2\nabla_{[l}\nabla_{m]} R^i_{jkh} \equiv -R^i_{lma}\, R^a_{jkh} + R^a_{lmj}\, R^i_{akh} + R^a_{lmk}\, R^i_{jah} + R^a_{lmh}\, R^i_{jka} = 0, \quad (6.1)$$

the space V_n has been called *semi-symmetric Riemannian space* which induces a semi-symmetric metric on the space [158], [219]. The operator ∇_l stands for the covariant differentiation with respect to x^l and the extreme left member denotes the skew-symmetric part of $\nabla_l \nabla_m$.

§ 7. Finsler metric

Let R_n be an n-dimensional real vector space completely covered by a curvilinear coordinate system (x^i), $i = 1, 2, \ldots, n$. Let the coordinate system (x^i) transform onto another coordinate system $(\overline{x}^a)$, $a = 1, 2, \ldots, n$ such that the functions

$$\overline{x}^a = \overline{x}^a(x^1, x^2, \ldots, x^n) \tag{7.1}$$

are at least of class C^2 (i.e. differentiable with respect to their arguments x^i's up to second order) and the Jacobian $(\partial \overline{x}^a / \partial x^i)$ of the transformation is non-zero. Let the parametric equations

$$x^i = x^i(t) \tag{7.2}$$

define a curve C in R_n. The derivatives $\dot{x}^i = dx^i / dt$ form the components of the tangent vector to C at the point P (x^i). The $2n$-tuple $(x^1, x^2, \ldots, x^n, \dot{x}^1, \dot{x}^2, \ldots, \dot{x}^n)$, briefly denoted as $(x^i, \dot{x}^i)$, is called the *line-element* of C. For distinction purposes, x^i's are called the positional coordinates of the point P and $\dot{x}^i$'s indicate the directional coordinates at P. The point P is called the *centre of the line-element.* Either sets of variables (x^i) and $(\dot{x}^i)$ are linearly independent to each other. The $\dot{x}^i$'s span a tangent manifold denoted by T_n(P) at P.

We now consider a function F of the line-element defining the infinitesimal (arcual) distance ds between the points P (x^i) and Q $(x^i + dx^i)$:

$$ds = F(x^i, dx^i), \tag{7.3}$$

and satisfying the conditions:

Condition 7.1. The function F is of class at least C^7 in all its $2n$ arguments.

Condition 7.2. The function F is positively homogeneous of degree 1 in its directional arguments $\dot{x}^i$'s, that is

$$F(x^i, k\dot{x}^i) = k\,F(x^i, \dot{x}^i), \quad \text{with} \quad k > 0, \tag{7.4}$$

or, by Euler's theorem,

$$\dot{x}^i\, \dot{\partial}_i F(x^i, \dot{x}^i) = F(x^i, \dot{x}^i), \qquad (\dot{\partial}_i \equiv \partial / \partial \dot{x}^i). \tag{7.5}$$

Condition 7.3. The function F is positive unless all its variables $\dot{x}^i$'s vanish simultaneously, that is

$$F(x^i, \dot{x}^i) > 0 \qquad \text{with} \qquad \sum_i (\dot{x}^i)^2 \neq 0. \tag{7.6}$$

Condition 7.4. The quadratic form $(\dot{\partial}_i \dot{\partial}_j F^2)\, \xi^i \xi^j$ is positive definite for all the variables ξ^i's.

Definition 7.1. The space R_n with the function F satisfying above conditions is called an n-dimensional *Finsler manifold.* The function F is called the *metric* or *fundamental function* of the manifold.

For Eq. (7.4), the function F^2 and its derivative $\dot{\partial}_j F^2$ are positively homogeneous of degrees 2 and 1 respectively in the variables $\dot{x}^i$'s. Hence, there hold the relations

$$\dot{x}^j\, \dot{\partial}_j F^2 = 2F^2 \quad \text{and} \quad \dot{x}^i\, \dot{\partial}_i(\dot{\partial}_j F^2) = \dot{\partial}_j F^2, \tag{7.7}$$

by Euler's theorem on homogeneous functions. Now, we define the functions

$$g_{ij}(x^k, \dot{x}^k) \equiv (1/2)\, \dot{\partial}_i\, \dot{\partial}_j\, F^2, \tag{7.8}$$

implying

$$g_{ij}(x^k, \dot{x}^k)\, \dot{x}^i\, \dot{x}^j = (1/2)\, \dot{x}^j \{\dot{x}^i\, \dot{\partial}_i(\dot{\partial}_j F^2)\} = F^2(x^k, \dot{x}^k), \tag{7.9a}$$

by Eq. (7.7). Applying Eq. (7.4) again and noting the homogeneity of g_{ij} (degree zero in $\dot{x}^k$'s) and F^2, we have

$$g_{ij}(x^k, dx^k) \equiv g_{ij}(x^k, \dot{x}^k dt) = (dt)^0\, g_{ij}(x^k, \dot{x}^k);$$

and

$$F^2(x^k, dx^k) = F^2(x^k, \dot{x}^k dt) = (dt)^2 F^2(x^k, \dot{x}^k).$$

Accordingly, Eq. (7.9a) reduces to

$$g_{ij}(x^k, dx^k)\, dx^i\, dx^j = F^2(x^k, dx^k) = (ds)^2, \tag{7.9b}$$

by Eq. (7.3). Above equation gives a *Finslerian metric*. The space R_n equipped with such a metric is called a *Finsler space* and will be denoted by F_n.

§ 8. Generalized Finslerian metric

Analogous to a metric discussed in § 5, a *generalized Finslerian metric* has been considered by Shamihoke [197] and others, where the metric coefficients $g_{ij}(x^k, \dot{x}^k)$ are not necessarily symmetric in i, j.

§ 9. Berwald metric

Defining a tensor

$$C_{ijk}(x^h, \dot{x}^h) \equiv (1/2)\, \dot{\partial}_i\, g_{jk} = (1/4)\, \dot{\partial}_i\, \dot{\partial}_j\, \dot{\partial}_k\, F^2, \qquad (9.1)$$

in continuation with the notation of § 7 and considering its vanishing covariant derivative with respect to x^h:

$$\nabla_h C_{ijk} = 0, \qquad (9.2)$$

L. Berwald calls a Finslerian metric as *affinely connected* if there holds above condition. Later authors renamed it after Berwald. The operator ∇_h denotes the covariant derivation with respect to positional coordinates x^h in the sense of Cartan (cf. Eq. (4.5.14) below).

§ 10. Landsberg metric

A Finslerian metric in Eq. (7.9b) satisfying

$$l_i\, G^i_{jkh} = 0, \qquad (10.1)$$

is called a *Landsberg metric* [83]. The covariant vector field $l_i(x^j, \dot{x}^j)$ is the unit vector along the element of support $\dot{x}^i$:

$$l_i = g_{ij}\, l^j = g_{ij}\, \dot{x}^j / F, \qquad (10.2)$$

and G^i_{jkh} is a mixed tensor defined as the directional derivative of the Berwald's connection parameters G^i_{kh}:

$$G^i_{jkh} \equiv \dot{\partial}_j\, G^i_{kh}. \qquad (10.3)$$

This tensor is symmetric in its all lower indices. Numata [159] establishes a comparison between Landsberg and Riemannian spaces and proves the:

Theorem 10.1. A Landsberg space F_n, $n > 2$, of scalar curvature is Riemannian with non-zero constant Riemannian curvature.

The connection parameters G^i_{kh} are independent of the directional arguments in an affinely connected Finsler space, thus, satisfying

$$G^i_{jkh} = 0. \tag{10.4}$$

Therefore, the condition vide Eq. (10.1) is identically satisfied in an affinely connected Finsler space leading to:

Example 10.1. A Berwald space is a Landsberg space.

§ 11. Randers' metric

Lichnerowicz [67] discusses a Finsler metric for electromagnetic field by decomposing the function

$$F(x^i, \dot{x}^i) = \sqrt{(a_{ij}\, \dot{x}^i\, \dot{x}^j)} + b_i\, \dot{x}^i, \tag{11.1}$$

where a_{ij} form the Riemannian fundamental tensor and b_i form a non-zero covariant vector. Both of these are functions of positional coordinates only. Such a metric has been called after Randers [202], [240]. Numata [160] discusses the torsion tensors of a Finsler space with such a metric.

§ 12. *C*-reducible (Finsler) metric

Matsumoto [74], [75] defines the angular metric tensor h_{ij}:

$$h_{ij}(x^k, \dot{x}^k) = g_{ij} - l_i\, l_j, \tag{12.1}$$

where the functions g_{ij} are given by Eq. (7.8) and the unit vector l_i by Eq. (10.2). If the torsion tensor C_{ijk} defined by Eq. (9.1) satisfies

$$C_{ijk} = (h_{ij}\, C_k + h_{jk}\, C_i + h_{ki}\, C_j)\,/\,(n+1) \tag{12.2}$$

for the vector

$$C_k \equiv C_{ijk}\, g^{ij}, \qquad (12.3)$$

obtained on transvection of C_{ijk} by the associate metric tensor g^{ij}, the metric of Finsler space F_n, $n \geq 3$, has been called *C-reducible* by Matsumoto [75].

§ 13. *P*-reducible (Finsler) metric

Matsumoto [80] also introduces a *P*-reducible (Finsler) metric by taking the second curvature tensor $P_{ijkh} \equiv P^r_{ikh}\, g_{rj}$ of Cartan in the form

$$P_{ijkh}\, \dot{x}^i \equiv P_{jkh} = (h_{jk}\, P_h + h_{kh}\, P_j + h_{hj}\, P_k) \,/\, (n+1) \qquad (13.1)$$

where

$$P_j \equiv P^k_{jk} = P_{jkh}\, g^{kh}. \qquad (13.2)$$

Shibata [200] calls the tensor P_{jkh} as the (v) hv-torsion tensor and proves the:

Theorem 13.1. A *P*-reducible (Finsler) space F_n, $n \geq 3$, of non-zero scalar curvature is *C*-reducible.

§ 14. P_2-like (Finsler) metric

When the second curvature tensor $P_{ijkh}\,(x^l, \dot{x}^l)$ of Cartan (cf. Rund [189]) satisfies

$$P_{ijkh} = P_i\, C_{jkh} - P_j\, C_{ikh} \qquad (14.1)$$

for some covariant vector P_i and the torsion tensor C_{ijk} defined by Eq. (9.1), the metric of a Finsler space F_n, $n \geq 3$, has been called "P_2-like" by Matsumoto [74]. Deriving necessary consequences of such a metric he proves the:

Theorem 14.1. A P_2-like F_n, implies vanishing of Cartan's either second curvature tensor P_{ijkh} or his first curvature tensor $S_{ijkh} \equiv S^r_{ikh}\, g_{rj}$.

He further establishes that the first alternative causes vanishing of the covariant derivative of S_{ijkh}:

$$P_{ijkh} = 0 \quad \Rightarrow \quad S_{ijkh|l} = 0, \qquad (14.2)$$

where the symbol $_{|l}$ denotes the covariant derivative with respect to x^l in the sense of Cartan (cf. Rund [189]).

§ 15. S_3-like (Finsler) metric

If there exists a scalar function S so that the first curvature tensor S_{ijkh} of Cartan satisfies

$$F^2 S_{ijkh} = S\,(h_{ik}\,h_{jh} - h_{ih}\,h_{jk}), \qquad (15.1)$$

where the tensor h_{ij} is given by Eq. (12.1), the metric of a Finsler space F_n, $n \geq 4$, has been called "S_3-like" by Matsumoto [74], [75]. He [74], further, establishes independence of the function S on the directional coordinates $\dot{x}^{i}$'s for such metrics.

§ 16. Kropina metric

Kropina [62] considers a Finsler space where the fundamental function $F(x^i, \dot{x}^i)$ is of the form

$$F = (a_{ij}\ \dot{x}^i\ \dot{x}^j)\,/\,b_k\ \dot{x}^k, \qquad (16.1)$$

where a_{ij} and b_k are explained in § 11. Such a metric is called after him.

Matsumoto [75] proves that the spaces with metrics given by either Eq. (11.1) or Eq. (16.1) are both C-reducible. Shibata [201] studied geometry of Finsler spaces with Kropina metric.

§ 17. Kawaguchi metric

Let the derivatives of the positional coordinates x^i's with respect to the parameter t up to any order, say p, be denoted by $x^{(p)i}$. Considering a function F of arguments x^i's, $\dot{x}^i$'s, $\ddot{x}^i$'s, … , $x^{(p)\,i}$'s, Akitsugu Kawaguchi defined the infinitesimal distance ds between two neighbouring points P (x^i) and Q $(x^i + dx^i)$ by the pth positive root of the function F:

$$ds = [F\{x^i,\ \dot{x}^i,\ \ddot{x}^i,\ \ldots,\ x^{(p)\,i}\}]^{1/p}. \qquad (17.1)$$

Such a metric has been called as *Kawaguchi metric* [56], [57]. A. Kawaguchi [56] also defined a special metric by considering a function F of special form:

$$F = A_i\,(x^j,\ \dot{x}^j)\ \ddot{x}^i + B\,(x^j,\ \dot{x}^j), \qquad (17.2)$$

where A_i is a covariant vector and B is a scalar function.

§ 18. Areal metric

The concept of linear and arcual metrics has been further generalized. Instead of measuring the arc-length of a curve the infinitesimal area of a region (of a space) has been defined by a function. Let the n coordinates x^i's define an m-dimensional sub-manifold V_m of a V_n:

$$V_m:\ x^i = x^i(u^\alpha), \qquad \alpha = 1, 2, \ldots, m < n. \tag{18.1}$$

Let the derivatives of x^i's with respect to the coordinates u^α's (of V_m) be considered up to m^{th} order. Taking a function

$$F\,\{x^i, \partial x^i/\partial u^\alpha, \partial^2 x^i/(\partial u^\alpha)^2, \ldots, \partial^m x^i/(\partial u^\alpha)^m\} \tag{18.2}$$

of x^i's and their derivatives Brickell [24] defines the metric

$$I = \int F\, du^1\, du^2 \ldots du^m, \tag{18.3}$$

and calls it the *areal metric*. Al-Borney [4] discussed the angular metric in areal spaces whereas Davies [35] defined a Euclidean connection in such spaces.

§ 19. Kähler metric

Consider a complex coordinate system (z^i) together with their complex conjugate system $(\bar{z}^i)$:

$$z^i = x^i + \sqrt{(-1)}\,y^i \qquad \text{and} \qquad \bar{z}^i = x^i - \sqrt{(-1)}\,y^i, \tag{19.1}$$

where both (x^i) and (y^i) form real coordinate systems in an n-dimensional space. Taking the metric coefficients g_{ij} as functions of the $2n$ complex coordinates $(z^i, \bar{z}^i)$ and imposing certain conditions the metrics of complex and almost complex spaces have been studied [236]. Metrics of Kähler, almost Kähler, Tachibana are few of them in the geometry of complex and almost complex spaces.

CHAPTER 4

BASIC CONCEPTS OF FINSLERIAN GEOMETRY

This chapter deals with some aspects of the fundamental theory of Finsler manifolds with special emphasis on the covariant differentiation related to the Euclidean connection of Cartan and Berwald.

§ 1. Finsler manifold

Let R be some region of an n-dimensional real space R_n covered completely by a coordinate system such that any point P ε R is represented by a set of n independent real variables x^i, $i = 1, 2, \dots, n$, called the coordinates of P. A transformation of these coordinates is represented by a set of n equations

$$x^{i'} = x^{i'}(x^1, x^2, \dots, x^n), \qquad i', j', \dots = 1, 2, \dots, n; \qquad (1.1)$$

where $(x^{i'})$ represent the coordinates of P in new variables. We assume that the functions $x^{i'}$ are at least of class C^2 and the Jacobian of the transformation vide Eqs. (1.1) is non-zero, that is

$$\det\,(\partial_i\, x^{i'}) \neq 0, \qquad (\partial_i \equiv \partial / \partial x^i). \qquad (1.2)$$

Definition 1.1. A set of points of R whose coordinates are expressible as functions of a single parameter, say t, is regarded as a *curve* of R_n.

Thus, the parametric equations

$$x^i = x^i(t) \qquad (1.3)$$

represent a curve C of R_n. The derivatives of the functions x^i with respect to the parameter t :

$$\dot{x}^i = d\,x^i / d\,t, \qquad (1.4)$$

constitute the components of a vector tangential to C at the point P. The combination $(x^i, \dot{x}^i)$ is called a *line-element* of C. To distinguish two types of arguments in the line-element we call x^i's the positional coordinate and $\dot{x}^i$'s the directional coordinates [3)] of P. The point P (x^i) is called the *centre* of the line-element.

As seen in the previous chapter (§ 7), a function F of line-elements satisfying certain conditions defines the metric of the space called a *Finsler space*.

§ 2. The metric tensor

A set of quantities $g_{ij}(x^k, \dot{x}^k)$ [4] defined by Eq. (3.7.8) and

$$g_{ij}\, g^{jk} = \delta_i^k = 1 \text{ (if } i = k), \quad 0 \text{ (if } i \neq k), \tag{2.1}$$

constitute the components of a symmetric tensor **g** called the *metric tensor* of the manifold. According to the condition 3.7.2, g_{ij} are positively homogeneous of degree 0 in $\dot{x}^i$'s and, therefore, they satisfy the following identities

$$y_i \equiv g_{ij}\, \dot{x}^j = (1/2)\, \dot{\partial}_i F^2 = F\, \dot{\partial}_i F, \tag{2.2}$$

and

$$y_i\, \dot{x}^i = g_{ij}\, \dot{x}^i\, \dot{x}^j = F^2. \tag{2.3}$$

From Eq. (1.4) it follows that the magnitude of the vector [5] $\dot{x}^i$ is the scalar function F. For the unit vector l^i in the direction of $\dot{x}^i$ given by Eq. (3.10.2), the Eqs. (2.2) and (2.3) yield

$$l_i \equiv g_{ij}\, l^j = (y_i / F) = \dot{\partial}_i F \quad (2.4); \qquad l^i\, l_i = g_{ij}\, l^i\, l^j = 1 \quad (2.5)$$

respectively. For Condition 3.7.2, Eq. (2.3) may also determine an important relation connecting the metric tensor with the function F:

$$F(x^k, dx^k) = \sqrt{\{g_{ij}(x^k, dx^k)\, dx^i\, dx^j\}}. \tag{2.6}$$

This relation gives rise to the following particular cases.

Case 2.1. If g_{ij} are independent of the directional arguments the formula (2.6) determines a *Riemannian metric* and the manifold with such a metric becomes Riemannian.

Case 2.2. If g_{ij} are independent [6] of the positional coordinates the formula (2.6) determines metric of a *Minkowskian manifold.*

Thus, as a Riemannian manifold is locally Euclidean, a Finsler manifold is locally Minkowskian.

Lemma 1.1. The directional derivatives of the vectors y_k, l^i and l_j are evaluated as [189]:

$$\text{a)}\ \dot{\partial}_j y_k = g_{jk}, \quad \text{b)}\ \dot{\partial}_j l^i = (\delta^i_j - l^i l_j)/F, \tag{2.7}$$

and

$$\dot{\partial}_j l_k = g_{ak}(\delta^a_j - l^a l_j)/F = (g_{jk} - l_j l_k)/F. \tag{2.8}$$

The tensor C_{ijk}, defined by Eq. (3.9.1), is positively homogeneous of degree – 1 in its directional arguments. Being symmetric in all its indices, it satisfies

$$C_{ijk}\,\dot{x}^i = C_{ijk}\,\dot{x}^j = C_{ijk}\,\dot{x}^k = 0; \tag{2.9}$$

and

$$(\partial_h C_{ijk})\,\dot{x}^i = (\partial_h C_{ijk})\,\dot{x}^j = (\partial_h C_{ijk})\,\dot{x}^k = 0; \tag{2.10}$$

Putting

$$A_{ijk} \equiv F\,C_{ijk}, \quad C^h_{ik} = g^{hj}\,C_{ijk}, \tag{2.11}$$

and

$$A^h_{ik} = g^{hi}\,A_{ijk} = F\,C^h_{ik}; \tag{2.12}$$

we get a new set of tensor fields found useful in our later discussion. These tensors are also symmetric in their lower indices and are positively homogeneous of degree 0, – 1, 0 respectively. Analogous to the identities (2.9) we have

$$A_{ijk}\,\dot{x}^i = A_{ijk}\,\dot{x}^j = A_{ijk}\,\dot{x}^k = 0, \tag{2.13}$$

and

$$C^h_{ik}\,\dot{x}^i = C^h_{ik}\,\dot{x}^k = A^h_{ik}\,\dot{x}^i = A^h_{ik}\,\dot{x}^k = 0. \tag{2.14}$$

Matsumoto [73], p. 75 calls the tensors C_{ijk} and C^i_{jk} the *torsion tensors*, whereas Moór [150] uses the same terminology for the tensors A_{ijk} and A^h_{ik}. However, such specific names for these tensors are neither used by Rund [189] nor by Cartan [33].

§ 3. Generalized Christoffel symbols

Like Riemannian geometry, we define the generalized Christoffel symbols of the first kind

$$\gamma_{ikj} \equiv \partial_{(i}\,g_{j)k} - (1/2)\,\partial_k\,g_{ij}.\,^{7)} \tag{3.1}$$

These symbols are symmetric in their first and last indices and are positively homogeneous of degree 0 in $\dot{x}^i$'s. Introducing the second kind symbols

$$\gamma^i_{jk} \equiv g^{ih}\,\gamma_{jhk}, \tag{3.2}$$

the differential equations of the geodesics may be obtained (cf. [189], pp. 50-52):

$$\ddot{x}^h + \gamma^h_{ij}\,\dot{x}^i\,\dot{x}^j = (\ddot{s}/\dot{s})\,\dot{x}^h; \; x''^h + \gamma^h_{ij}\,x'^i\,x'^j = 0, \tag{3.3}$$

where a dot ($\cdot$) respectively dash ($'$) above x denotes the differentiation with regard to the parameter t (resp. the arc-length s of the curve).

§ 4. Two processes of Rund's covariant differentiation

Like Riemannian geometry, the partial derivatives of an object [8)] with respect to the positional coordinates do not, in general, preserve its vector or tensor character. But, the addition or subtraction of certain suitable terms involving the functions called the *connection parameters* to the partial derivatives fulfills the above requirement. Such an augmentation of the partial derivation is called the *covariant differentiation* with respect to the concerned positional coordinates. In Riemannian geometry, there exists a unique connection formed by second kind Christoffel symbols giving rise to the process of covariant differentiation. On the other hand, in Finsler geometry, there exist a number of connections given by Rund, Cartan and Berwald for which the corresponding processes of covariant derivation of an object may be defined preserving its vector or tensor character. However, the generalized Christoffel symbols in Finsler geometry fail to form a connection for which the covariant derivation of an object, when defined may not preserve its vector or tensor character [9)]. In the present section we take up Rund's connection parameters and the corresponding processes of covariant derivation only postponing the discussion of other connection parameters and the corresponding covariant derivation processes to the later sections.

Firstly, when an object is a function of a single parameter, say t, and secondly, when it is a function of the line-elements, Rund [186] - [189] introduces two types of connection parameters:

$$P^i_{jk} \equiv \gamma^i_{jk} - C^i_{jh}\,\gamma^h_{km}\,\dot{x}^m, \tag{4.1}$$

and

$$P^{*i}_{jk} \equiv \gamma^{i}_{jk} - \{2\, C^{i}_{h(j}\, P^{h}_{k)m} - g^{il}\, C_{jkh}\, P^{h}_{lm}\}\, \dot{x}^{m}. \qquad (4.2)$$

The first kind parameters are not symmetric but those of the second kind are symmetric in their lower indices. For Eqs. (2.9), (2.13) and (2.14), they are connected by

$$P^{i}_{jk}\, \dot{x}^{j} \equiv \gamma^{i}_{jk}\, \dot{x}^{j} \qquad (4.3); \qquad P^{i}_{jk}\, \dot{x}^{k} = P^{*i}_{jk}\, \dot{x}^{k} \qquad (4.4)$$

and therefore,

$$P^{*i}_{jk}\, \dot{x}^{j}\, \dot{x}^{k} = P^{i}_{jk}\, \dot{x}^{j}\, \dot{x}^{k} = \gamma^{i}_{jk}\, \dot{x}^{j}\, \dot{x}^{k}. \qquad (4.5)$$

For the connection parameters P^{i}_{jk}, Rund defines the derivative

$$\delta X^{i} / \delta t = dX^{i} / dt + X^{j} P^{i}_{jk} \dot{x}^{k} \qquad (4.6)$$

of a vector field $X^{i}(t)$ and calls it the δ-*derivative*. This process preserves the vector character of $X^{i}(t)$.

Secondly, for the vector field X^{i} depending upon the line-elements (x^{j}, ξ^{j}) [10], Rund defines the *partial* δ-*derivative* of $X^{i}(x^{j}, \xi^{j})$ with respect to x^{k} for the connection parameters P^{*i}_{jk} by

$$\Re_{k}\, X^{i} = \partial_{k} X^{i} + (\dot{\partial}_{j} X^{i})(\partial_{k} \xi^{j}) + X^{j}\, P^{*i}_{jk}. \text{ [11]} \qquad (4.7)$$

The entities $\Re_{k} X^{i}$ constitute a mixed tensor of type (1, 1). Hereafter, we shall refer this tensor as the *Rund's covariant derivative* of the vector $X^{i}(x^{j}, \xi^{j})$ with respect to x^{k}. The connection parameters P^{*i}_{jk} may be called as the *Rund's connection parameters*. It is seen that this connection is, in general, a non-metric connection, that is, the covariant derivative of the metric tensor $g_{ij}(x^{k}, \xi^{k})$ does not, in general, vanish ([189], p. 61). In fact, it is given by

$$\Re_{k}\, g_{ij} = 2C_{ijh}\, \Re_{k}\, \xi^{h}. \qquad (4.8)$$

This equation represents a generalization of the Ricci Lemma of Riemannian geometry where the tensor C_{ijk} vanishes identically making the connection metric.

§ 5. Euclidean connection of Cartan and his two processes of covariant derivation

As seen in the preceding section, the Ricci Lemma of Riemannian geometry does not hold in Finsler manifolds regarded as locally Minkowskian. This divergence affects the radical differences between the further developments of the theories of Finsler and Riemannian manifolds as the Ricci Lemma plays a vital role in the latter case. However, this divergence could be overcome by the introduction of so-called "*Euclidean connection*" of Cartan [32], [33]. Thus, the preservation of Ricci Lemma was once more ensured. Cartan treats the theory of Finsler manifolds from an entirely different point of view. His approach to Finsler geometry exerted the great influence upon the general development of the subject. Several mathematicians expressed the opinion that the theory had, thus, attained its final form. However, this was not altogether correct as we shall see in the next section, while introducing the Berwald's theory of Finsler manifolds. Without going into detailed account of Cartan's theory [12)] we shall give here the formalism of his connection parameters and his two processes of covariant derivation.

Taking a vector field X^i Cartan considers its variation under an infinitesimal change of its line-element [13)] $(x^i, \dot{x}^i)$ to $(x^i + dx^i, \dot{x}^i + d\dot{x}^i)$ by the covariant differential:

$$D_k X^i = dX^i + X^i(\Gamma^i_{jk}\, dx^k + C^i_{jk}\, d\dot{x}^k). \qquad (5.1)$$

where the functions Γ^i_{jk} and C^i_{jk} are restricted by the additional postulates. It is seen that the functions C^i_{jk}, introduced by Cartan, coincide with those defined by Eq. (3.9.1) together with Eq. (2.11). But, the functions Γ^i_{jk} are seen to satisfy ([189]. Eq. (3.1.30))

$$\Gamma^i_{jk} = \gamma^i_{jk} + C_{jkh}\, g^{il}\, \dot{\partial}_l\, G^h - C^i_{kh}\, \dot{\partial}_l\, G^h, \qquad (5.2)$$

where

$$2G^i = \gamma^i_{jk} \dot{x}^j \dot{x}^k. \qquad (5.3)$$

The functions Γ^i_{jk}, being not symmetric, fail to provide the Euclidean connection of Cartan. To overcome this difficulty an additional postulate is taken and a new set of parameters Γ^{*i}_{jk} is defined:

$$\Gamma^{*i}_{jk} = \Gamma^{i}_{jk} - C^{i}_{jh}\,\dot{\partial}_k G^h. \tag{5.4}$$

It may be noticed that these functions are symmetric and provide the Euclidean connection of Cartan.

Using Eq. (2.14) it may be easily verified that the two types of parameters introduced by Cartan are also connected by

$$\Gamma^{i}_{jk}\,\dot{x}^j = \Gamma^{*i}_{jk}\,\dot{x}^j. \tag{5.5}$$

It was established by Davies that Rund's connection parameters $P^{*i}_{\ jk}$ coincide with Γ^{*i}_{jk}. However, the construction of these connection parameters is quite different. For Eqs. (2.9), (2.14) and (5.3), it may be deduced from Eqs. (4.1) and (5.2) that the parameters P^{i}_{jk} and Γ^{i}_{jk} are connected by

$$P^{i}_{jk}\,\dot{x}^k = \Gamma^{\ i}_{k\ j}\,\dot{x}^k. \tag{5.6}$$

Consequently, from Eqs. (4.4) and (5.5) we have

$$P^{i}_{jk}\,\dot{x}^k = P^{*i}_{jk}\,\dot{x}^k = \Gamma^{*i}_{k\ j}\,\dot{x}^k = \Gamma^{\ i}_{k\ j}\,\dot{x}^k. \tag{5.7}$$

At this stage it may be worthwhile to find relations connecting the derivatives $G^i_h \equiv \dot{\partial}_h\, G^i$ with various types of connection parameters discussed above. Differentiating the functions given by Eq. (5.3) with respect to $\dot{x}^h$ and using Eq. (3.2), we get

$$2G^i_h = (\dot{\partial}_h\, g^{il})\,\gamma_{jlk}\,\dot{x}^j\,\dot{x}^k + g^{il}(\dot{\partial}_h\gamma_{jlk})\,\dot{x}^j\,\dot{x}^k + 2\gamma^{\ i}_{h\ k}\,\dot{x}^k.$$

Using Eqs. (3.9.1), (2.10) and (3.1) it may be seen that the middle term in the second member of above identity vanishes identically. Also, for Eq. (3.2), the first term therein simplifies as follows:

$$(\dot{\partial}_h g^{il})\,g_{lm}\gamma^{m}_{jk}\dot{x}^j\dot{x}^k = -\,g^{il}(\dot{\partial}_h g_{lm})\gamma^{m}_{jk}\dot{x}^j\dot{x}^k = -2C^{i}_{hm}\gamma^{m}_{jk}\dot{x}^j\dot{x}^k,$$

where Eqs. (3.9.1), (2.1) and (2.12) are also used. Consequently, we have

$$G^i_h \equiv (\gamma^{\ i}_{h\ k} - C^{i}_{hm}\,\gamma^{m}_{jk}\,\dot{x}^j)\,\dot{x}^k, \tag{5.8}$$

or, in consequence of Eq. (4.1),

$$G^i_j = P^i_{jk}\dot{x}^k. \tag{5.9}$$

After this brief account of the Cartan's connection parameters we return to the original equation (5.1). For

$$dX^i = (\partial_k X^i)\,dx^k + (\dot{\partial}_k X^i)\,d\dot{x}^k,$$

Eq. (5.1) may be written as

$$DX^i = (\partial_k X^i + X^j \Gamma^i_{jk})\,dx^k + (\dot{\partial}_k X^i + X^j C^i_{jk})\,d\dot{x}^k. \tag{5.10}$$

In order to evaluate $d\dot{x}^k$ we apply the formula (5.1) to the unit vector l^k acting along $\dot{x}^k$, and use Eqs. (3.10.2) and (2.14):

$$d\dot{x}^k = F\,Dl^k + (dF/F)\,\dot{x}^k - \dot{x}^j \Gamma^k_{jh}\,dx^h. \tag{5.11}$$

Eliminating $d\dot{x}^k$ from Eqs. (5.10) and (5.11), and simplifying by means of Eqs. (2.14), (5.4), (5.7) and (5.9), we get

$$DX^i = (\dot{\nabla}_k X^i)\,F\,Dl^k + (\nabla_k X^i)\,dx^k + (dF/F)\,(\dot{x}^k \dot{\partial}_k X^i), \tag{5.12}$$

where we have put [14)]

$$\dot{\nabla}_k X^i \equiv \dot{\partial}_k X^i + X^j C^i_{jk}, \tag{5.13}$$

and

$$\nabla_k X^i \equiv \partial_k X^i - (\dot{\partial}_j X^i)\,G^j_k + X^j \Gamma^{*i}_{jk}. \tag{5.14}$$

The processes illustrated by Eqs. (5.13) and (5.14) may be called the processes of *Cartan's covariant derivation*. Either of these derivatives are the components of tensors of type (1, 1). In particular, these covariant derivatives vanish for the following quantities:

$$D\mathbf{g} = \dot{\nabla}_k \mathbf{g} = \nabla_k \mathbf{g} = 0, \tag{5.15}$$

$$\nabla_k \dot{x}^i = \nabla_k l^i = \nabla_k l_j = \nabla_k F = \nabla_k y_j = 0. \tag{5.16}$$

As remarked earlier the connection parameters P^{*i}_{jk} and Γ^{*i}_{jk} are the same yet the corresponding processes of covariant derivations are not always same. For a vector field $X^i(x^j)$ depending upon the positional coordinates only, so that $\dot{\partial}_j X^i = 0$, there follows immediately

$$\Re_k X^i = \nabla_k X^i$$

from Eqs. (4.7) and (5.14). However, a significant difference lies in the fact that the covariant differential DX^i, given by Eq. (5.12), depends upon the variation Dl^k of the unit vector l^k along the directional coordinate $\dot{x}^k$, whereas the absolute differential

$$\delta X^i \equiv (\Re_k X^i)\, dx^k \ ,$$

given in the previous section, depends only on the direction $\dot{x}^k$ along which the process of covariant derivation takes place. On the other hand, when the vector field X^i is a function of line-elements [15], it follows from Eqs. (4.7) and (5.14) that the two processes of covariant derivation are connected by

$$(\Re_k - \nabla_k) X^i = (\dot{\partial}_j X^i)(\partial_k \xi^j + G^j_k). \tag{5.17}$$

§ 6. Berwald's connection and his covariant derivation

Staring with the positively homogeneous functions G^i of degree 2 in their directional arguments, defined by Eq. (5.3), Berwald considers their partial derivatives:

$$G^i_{jk} = \dot{\partial}_j G^i_k = \dot{\partial}_j \dot{\partial}_k G^i . \tag{6.1}$$

As such, the functions G^i_{jk} are also positively homogeneous of degree 0 in $\dot{x}^i$'s, and satisfy

$$G^i_{jk} \dot{x}^j = G^i_k, \tag{6.2a}$$

together with

$$G^i_k \dot{x}^k = 2G^i . \tag{6.3}$$

For Eqs. (5.7) and (5.9), the functions G^i_{jk} can be also seen connected with the Cartan's connection parameters Γ^{*i}_{jk}:

$$G^i_k = G^i_{jk}\, \dot{x}^j = \Gamma^{*i}_{jk}\, \dot{x}^j, \tag{6.2b}$$

and [16]

$$G^i_{jk} - \Gamma^{*i}_{jk} = \dot{x}^h \dot{\partial}_j \Gamma^{*i}_{hk} = \nabla C^i_{jk}, \qquad (\nabla \equiv \dot{x}^h \nabla_h). \quad (6.4)$$

∇C^i_{jk}, being a tensor, the functions G^i_{jk} satisfy the same coordinate transformation law as for Γ^{*i}_{jk}. Thus, G^i_{jk} are the new connection parameters introduced for the first time by Noether [157], but used extensively by Berwald [18] - [21], hence known after Berwald.

Finsler spaces for which the connection parameters G^i_{jk} are independent of the directional arguments are called, by Berwald, "*affinely-connected*" spaces while Vagner [227] calls such spaces as "*Berwald spaces*". It is shown that a necessary and sufficient condition for a Finsler space to be affinely connected is given by Eq. (3.9.2), or

$$\nabla_h C^i_{jk} = 0. \quad (6.5)$$

Consequently, for Eq. (6.4), the connection parameters of both Berwald and Cartan are equal in an affinely connected Finsler space.

The covariant derivation of a vector X^i with respect to x^k for the connection parameters G^i_{jk} is defined in similar way as that of Cartan's second process of covariant derivation illustrated by Eq. (5.14):

$$\mathfrak{B}_k X^i \equiv \partial_k X^i - (\dot{\partial}_j X^i) G^j_k + X^j G^i_{jk}.^{17)} \quad (6.6)$$

The quantities $\mathfrak{B}_k X^i$ constitute a tensor of type (1, 1) and are called *Berwald's covariant derivatives* of the vector X^i with respect to x^k. In particular, the metric function, the unit vector l^i along the vector $\dot{x}^i$, and the vector field $\dot{x}^i$ itself have vanishing covariant derivatives:

$$\text{(a)}\ \mathfrak{B}_k F = 0, \quad \text{(b)}\ \mathfrak{B}_k l^i = 0, \quad \text{(c)}\ \mathfrak{B}_k \dot{x}^i = 0. \quad (6.7)$$

On the other hand, Berwald's covariant derivative of the metric tensor is, in general, non-zero. It is obtained in the form ([189], p. 80):

$$\mathfrak{B}_k g_{ij} = -2\, \nabla C_{ijk}. \quad (6.8)$$

Thus, once again it was established that the Ricci Lemma of Riemannian geometry does not necessarily hold in Finsler geometry. However, it does hold in an affinely-connected Finsler space. Non-vanishing of the covariant derivative of the metric tensor makes the main diversion of the Berwald's theory to that of Cartan.

Transvecting Eq. (6.8) by unit vector l^j, and noting Eq. (6.7b) and (6.4), one may also derive

$$\mathfrak{B}_j\, l_i = -2l^j\, \dot{x}^a\, \nabla_a C_{ijk} = -2\dot{x}^a\, \nabla_a (l^i C_{ijk}) = 0, \qquad (6.9)$$

for Eqs. (2.9). Processes of covariant derivation and directional derivation being commutative for a scalar function, Eq. (6.9) may be also derived alternately by forming covariant derivative of Eq. (2.4) and putting from Eq. (6.7a):

$$\mathfrak{B}_j\, l_i = \mathfrak{B}_j\, \dot{\partial}_i\, F = \dot{\partial}_i\, \mathfrak{B}_j\, F = 0.$$

Also, for Eqs. (6.7a) and (6.9), there follows

$$\mathfrak{B}_j\, y_i = \mathfrak{B}_j\, (F\, l_i) = 0. \qquad (6.10)$$

§ 7. Curvature tensors

Like Riemannian geometry, the successive processes of covariant derivations are not, in general, commutative. In fact, their difference is represented by a commutation formula (called the *Ricci identity* in Riemannian geometry) giving rise to a tensor field called the *curvature tensor*. In the following, we endeavour to find various commutation formulae and curvature tensors arising from different processes of covariant derivation discussed in previous sections.

First, we consider Rund's covariant derivation illustrated by Eq. (4.7). Forming its further covariant derivative with respect to x^j (in the sense of Rund), and taking the skew-symmetric part of the derived equation with respect to the indices *k* and *h*, we obtain the commutation formula

$$2\mathfrak{R}_{[j}\, \mathfrak{R}_{k]}\, X^i = \widetilde{K}^i_{jkh} X^h, \qquad (7.1)$$

where we have put

$$\widetilde{K}^i_{jkh}(x^l, \xi^l) = 2\{\partial_{[j} P^{*i}_{k]h} + (\dot{\partial}_l P^{*i}_{h[k})(\partial_{j]}\xi^l) + P^{*i}_{l[j} P^{*l}_{k]h}\}. \quad (7.2)$$

The tensor $\widetilde{K}^i_{jkh}(x^l, \xi^l)$[18], called by Rund as "*relative curvature tensor*" may be referred to here onwards as *Rund's curvature tensor*.

Next, we derive three curvature tensors of Cartan. In the first step we start with the process of covariant differentiation illustrated by Eq. (5.13). Operatng it by $\dot{\nabla}_j$, and forming the skew - symmetric part of the derived equation with respect to the indices *j* and *k*, we obtain

$$2\dot{\nabla}_{[j}\,\dot{\nabla}_{k]}\,X^i = 2\{\dot{\partial}_{[j}\,C^i_{k]h} + C^i_{r[j}C^r_{k]h}\}X^h. \tag{7.3}$$

To evaluate the first term in the second member of above identity, we use Eqs. (3.9.1) and (2.12) successively:

$$\dot{\partial}_j C^i_{kh} = (\dot{\partial}_j g^{il})\,C_{klh} + g^{il}\,\dot{\partial}_j C_{klh} = (\dot{\partial}_j g^{il})\,g_{lr} C^r_{kh} + (1/2)\,g^{il}\,\dot{\partial}_j\dot{\partial}_k g_{lh}.$$

Also, differentiating $g^{il}g_{lr} = \delta^i_r$ with respect to $\dot{x}^j$, and using (3.9.1) and (2.12) again, we have

$$(\dot{\partial}_j g^{il})\,g_{lr} = -g^{il}\,\dot{\partial}_j g_{lr} = -2g^{il}C_{jlr} = -2C^i_{jr}.$$

Therefore [19],

$$2\dot{\partial}_{[j}C^i_{k]h} = -4C^i_{r[j}C^r_{k]h}. \tag{7.4}$$

Putting from Eq. (7.4) in (7.3), and writing

$$S^i_{jkh} = 2C^r_{h[j}C^i_{k]r}, \tag{7.5}$$

we finally get the commutation formula

$$2\dot{\nabla}_{[j}\,\dot{\nabla}_{k]}\,X^i = S^i_{jkh}X^h. \tag{7.6}$$

The tensor $S^i{}_{jkh}$ is the *first of Cartan's three curvature tensors.*

Secondly, for a commutation formula involving both processes of Cartan's covariant derivation, from Eqs. (5.13) and (5.14), we derive

$$(\dot{\nabla}_j\nabla_k - \nabla_k\dot{\nabla}_j)X^i = (\dot{\partial}_j\Gamma^{*i}_{kh} - \nabla_k C^i_{jh})X^h - C^h_{jk}\nabla_h X^i + (\Gamma^{*h}_{jk} - G^h_{jk})\,\dot{\partial}_h X^i.$$

Putting from Eq. (6.4) and eliminating $\dot{\partial}_h X^i$ by means of Eq. (5.13), this identity reduces to

$$(\dot{\nabla}_j\nabla_k - \nabla_k\dot{\nabla}_j)X^i = P^i_{jkh}X^h - C^h_{jk}\nabla_h X^i - (\nabla C^h_{jk})\dot{\nabla}_h X^i, \tag{7.7}$$

where we have put [20])

$$P^i_{jkh} = \dot{\partial}_j \Gamma^{*i}_{kh} - \nabla_k C^i_{jh} + (\nabla C^l_{jk}) C^i_{hl}. \qquad (7.8a)$$

This tensor is called the *second curvature tensor of Cartan*. A more useful form of this tensor is also given by ([189], Eq. (4.1.31))

$$P^i_{jkh} = \nabla_h C^i_{jk} - \nabla^i C_{jkh} + (\nabla C^i_{lj}) C^l_{kh} - (\nabla C^l_{hj}) C^i_{kl}, \qquad (7.8b)$$

where $\nabla^i \equiv g^{il} \nabla_l$.

Thirdly, evaluating the second order covariant derivative $\nabla_j \nabla_k X^i$ by means of Eq. (5.14), taking its skew-symmetric part with respect to the indices *j*, *k* and simplifying, we get

$$2\nabla_{[j} \nabla_{k]} X^i = K^i_{jkh} X^h - 2\{\partial_{[j} G^h_{k]} + (\dot{\partial}_l G^h_{j[}) G^l_{k]}\} \dot{\partial}_h X^i, \qquad (7.9)$$

where [21])

$$K^i_{jkh} = 2\{\partial_{[j} \Gamma^{*i}_{k]h} + (\dot{\partial}_l \Gamma^{*i}_{h[j}) G^l_{k]} + \Gamma^{*i}_{l[j} \Gamma^{*l}_{k]h}\}. \qquad (7.10)$$

Taking note of the relations (6.1) - (6.3) the coefficient of $\dot{\partial}_h X^i$ in Eq. (7.9) may be written as

$$2\{\partial_{[j} \Gamma^{*h}_{k]m} + (\dot{\partial}_l \Gamma^{*h}_{m[j}) G^l_{k]} + \Gamma^{*h}_{l[j} \Gamma^{*l}_{k]m}\} \dot{x}^m,$$

which, for Eq. (7.10), reduces to $K^h_{jkm} \dot{x}^m$. Thus, the commutation formula (7.9) simplifies to

$$2\nabla_{[j} \nabla_{k]} X^i = K^i_{jkh} X^h - (\dot{\partial}_h X^i) K^h_{jkm} \dot{x}^m. \qquad (7.11)$$

The tensor K^i_{jkm} is also a curvature tensor but Cartan reserves terminology "third curvature tensor" for a tensor yet to be introduced. Using Eq. (5.13), $\dot{\partial}_h X^i$ may be eliminated from Eq. (7.11), and the commutation formula, finally assumes the form

$$2\nabla_{[j} \nabla_{k]} X^i = R^i_{jkh} X^h - (\dot{\nabla}_h X^i) K^h_{jkm} \dot{x}^m, \qquad (7.12)$$

where we have put

$$R^i_{jkh} = K^i_{jkh} + C^i_{hl} K^l_{jkm} \dot{x}^m. \qquad (7.13)$$

The tensor R^{i}_{jkh} is called the *third curvature tensor of Cartan*.

While concluding this section, we also derive a commutation formula arising from Berwald's covariant derivation. Starting with the formula (6.6) and proceeding on the lines similar to those leading to the formula (7.11) we derive, after some simplification,

$$2\,\mathfrak{B}_{[j}\mathfrak{B}_{k]}X^{i} = H^{i}_{jkh}X^{h} - (\dot{\partial}_{h}X^{i})H^{h}_{jkm}\dot{x}^{m}, \tag{7.14}$$

where

$$H^{i}_{jkh} = 2\,\partial_{[j}G^{i}_{k]h} + (\dot{\partial}_{l}G^{i}_{h[j})G^{l}_{k]} + G^{i}_{l[j})G^{l}_{k]h}\} \tag{7.15}$$

is *Berwald's curvature tensor*.

§ 8. Some additional commutation formulae

A large number of commutation formulae have been obtained involving various processes of covariant differentiation and the partial derivation with respect to the directional coordinates $\dot{x}^{i}$'s [22)]. Some of these, found more useful in the subsequent discussion, are given below.

Theorem 8.1. Partial derivation with respect to $\dot{x}^{j}$ commutes with various processes of covariant derivation according to

$$(\dot{\partial}_{j}\dot{\nabla}_{k} - \dot{\nabla}_{k}\dot{\partial}_{j})X^{i} = (\dot{\partial}_{j}C^{i}_{kh})X^{h} + C^{h}_{jk}\dot{\partial}_{h}X^{i}, \tag{8.1}$$

$$(\dot{\partial}_{j}\nabla_{k} - \nabla_{k}\dot{\partial}_{j})X^{i} = (\dot{\partial}_{j}\Gamma^{*i}_{kh})X^{h} - (\nabla C^{h}_{jk})\,\dot{\partial}_{h}X^{i}, \tag{8.2}$$

$$(\dot{\partial}_{j}\mathfrak{B}_{k} - \mathfrak{B}_{k}\,\dot{\partial}_{j})\,X^{i} = G^{i}_{jkh}X^{h}, \tag{8.3}$$

where X^{i} is an arbitrary contravariant vector field.

Proof. Differentiating Eq. (5.13) with respect to $\dot{x}^{j}$, we get

$$\dot{\partial}_{j}\dot{\nabla}_{k}X^{i} = \dot{\partial}_{j}\dot{\partial}_{k}X^{i} + (\dot{\partial}_{j}C^{i}_{kh})X^{h} + C^{i}_{kh}\dot{\partial}_{j}X^{h}. \tag{8.4}$$

On the other hand, evaluating the covariant derivative of the tensor $\dot{\partial}_{j}X^{i}$ by means of the formula (5.13), we get

$$\dot{\nabla}_k \dot{\partial}_j X^i = \dot{\partial}_k \dot{\partial}_j X^i - C^h_{jk} \dot{\partial}_h X^i + C^i_{kh} \dot{\partial}_j X^h. \tag{8.5}$$

The commutation formula (8.1) follows immediately from Eqs. (8.4) and (8.5).

To establish formula (8.2), the derivatives indicated therein may be evaluated with the help of Eqs. (5.14) and (6.1). After cancellation of certain common terms we derive

$$(\dot{\partial}_j \nabla_k - \nabla_k \dot{\partial}_j) X^i = (\dot{\partial}_j \Gamma^{*i}_{kh}) X^h + (\Gamma^{*h}_{jk} - G^h_{jk}) \dot{\partial}_h X^i, \tag{8.6}$$

which, in consequence of Eq. (6.4), reduces to the form (8.2). The formula (8.3) follows immediately from Eq. (8.6) when we replace the process of covariant derivation and the corresponding connection parameters of Cartan by those of Berwald. It is to be noted that the last term in the second member of Eq. (8.6) then vanishes. However, an independent proof of formula (8.3) can be also given with the help of Eq. (6.6). //

§ 9. Various properties of curvature tensors

The curvature tensors in Finsler geometry satisfy many properties analogous to those of the Riemann curvature tensor in Riemannian geometry. In contrast to the Riemann curvature tensor the main diverse characteristics of the curvature tensors (other than Cartan's first curvature tensor S^i_{jkh}, which satisfies all the characteristics of Riemann curvature tensor, see Remark 9.1 below) in Finsler geometry are the following:

(i) The tensors analogous to Ricci tensor of Riemannian geometry are not, in general, symmetric in Finsler geometry.

(j) The associate curvature tensors defined by

$$\text{(a)}\ \Omega_{jkhl} = \Omega^i_{jkh}\, g_{il}, \qquad \text{(b)}\ \Omega_{jkhl}\, g^{il} = \Omega^i_{jkh}, \tag{9.1}$$ [23]

(Ω^i_{jkh} being a curvature tensor) are not symmetric pair-wise.

In the following, we derive some analytical properties of the curvature tensors. Firstly, it may be noticed that except the second curvature tensor of Cartan P^i_{jkh} rest all the curvature tensors and their associate curvature tensors defined by Eqs. (9.1) are skew-symmetric in their first two lower indices:

$$\text{(a)} \quad 2\Omega^{i}_{(jk)h} = 0, \qquad \text{(b)} \quad 2\Omega_{(jk)hl} = 0. \tag{9.2}$$

Secondly, it is interesting to note that the associate curvature tensors S_{jkhl}, P_{jkhl} and R_{jkhl} are also skew-symmetric in their last two indices like the associate Riemann curvature tensor in Riemannian geometry:

$$S_{jk(hl)} = P_{jk(hl)} = R_{jk(hl)} = 0. \tag{9.3}$$

For the first two of above results, we write an expression for the associate curvature tensors and use Eqs. (2.12) and (5.15) successively. The last result follows from the commutation formula (7.12), applied for the metric tensor g_{hl}, and using Eqs. (5.15). No other curvature tensors in Finsler geometry possess this property. For instance, the formula (7.11), when applied for the metric tensor g_{hl}, in consequence of Eqs. (3.9.1) and (5.15), yields

$$K_{jk(hl)} + C_{hlm} K^{m}_{jkr} \dot{x}^{r} = 0. \tag{9.4}$$

Thirdly, for pair-wise symmetry, we notice that S_{jkhl} is the only associate curvature tensor in Finsler geometry that satisfies

$$S_{jkhl} = S_{hljk}. \tag{9.5}$$

Fourthly, when transvected by the directional arguments the curvature tensors satisfy

$$S^{i}_{jkh} \dot{x}^{j} = S^{i}_{jkh} \dot{x}^{k} = S^{i}_{jkh} \dot{x}^{h} = 0, \tag{9.6}$$

$$\text{(a)} \quad P^{i}_{jkh} \dot{x}^{j} = P^{i}_{jkh} \dot{x}^{k} = 0, \qquad \text{(b)} \quad P^{i}_{jkh} \dot{x}^{h} = \nabla C^{i}_{jk}. \tag{9.7}$$

$$R^{i}_{jkh} \dot{x}^{h} = K^{i}_{jkh} \dot{x}^{h} = H^{i}_{jkh} \dot{x}^{h}. \tag{9.8a}$$

Also, transvection of the Eqs. (9.3) by g^{hl}, for Eq. (9.1b), yields

$$S^{i}_{jki} = P^{i}_{jki} = R^{i}_{jki} = 0. \tag{9.9}$$

Fifthly, defining tensors analogous to Ricci tensor of Riemannian geometry:

$$\Omega_{kh} \equiv \Omega^{i}_{ikh}, \tag{9.10}$$

we observe that S^i_{jkh} is the only curvature tensor that yields a symmetric tensor:

$$S_{kh} \equiv S^i_{ikh} = S_{hk}. \tag{9.11}$$

Because of the identities vide Eqs. (9.3), (9.5), (9.9) and (9.11), we have the:

Remark 9.1. The tensor S^i_{jkh} is the only curvature tensor in Finsler geometry which possesses almost all the characteristics of the Riemann curvature tensor of Riemannian geometry.

In the rest of this section, we consider the properties exclusively satisfied by Berwald's curvature tensor. Because of Eq. (6.1), the tensor field defined by

$$G^i_{lhj} = \dot{\partial}_l\, G^i_{hj} \tag{9.12}$$

is symmetric in all its lower indices and, therefore, the expression for Berwald's curvature tensor given by Eq. (7.15) may be rewritten as

$$H^i_{jkh} = 2\,\dot{\partial}_h\{\partial_{[j}G^i_{k]} + G^i_{l[j}G^l_{k]}\}. \tag{9.13}$$

Thus, putting

$$H^i_{jk} = 2\{\partial_{[j}G^i_{k]} + G^i_{l[j}G^l_{k]}\}, \tag{9.14}$$

the curvature tensor satisfies

$$\text{(a)}\quad H^i_{jkh} = \dot{\partial}_h H^i_{jk}, \qquad \text{(b)}\quad H^i_{jkh}\,\dot{x}^h = H^i_{jk}. \tag{9.15}$$

The last relation justifies the tensor character of the functions defined by Eq. (9.14), which are skew-symmetric in their lower indices and are positively homogeneous of degree 1 in $\dot{x}^i$'s. Hence, the identity (9.8a) may be rewritten as

$$R^i_{jkh}\,\dot{x}^h = K^i_{jkh}\,\dot{x}^h = H^i_{jk}. \tag{9.8b}$$

Further, defining a tensor H^i_j, called *deviation tensor* by Berwald:

$$H^i_j \equiv H^i_{jk}\dot{x}^k = 2\partial_j\, G^i - \dot{x}^k\partial_k\, G^i_j + 2G^i_{lj}G^l - G^i_l G^l_j, \tag{9.16}$$

the following identities can be also deduced:

$$H^i_{jk} = (2/3)\,\dot{\partial}_{[k}\, H^i_{j]} \quad (9.17); \qquad H^i_j\, \dot{x}^j = 0. \quad (9.18)$$

Clearly, the tensor H^i_j is positively homogeneous of degree 2 in $\dot{x}^i$'s. In view of Eqs. (9.8b) and (9.16) it bears close relationship with the curvature tensors K^i_{jkh}, R^i_{jkh} and H^i_{jkh}:

$$R^i_{jkh}\, \dot{x}^k \dot{x}^h = K^i_{jkh}\, \dot{x}^k \dot{x}^h = H^i_{jkh} \dot{x}^k \dot{x}^h = H^i_j\,. \quad (9.19)$$

Introducing the contractions:

$$\text{(a) } H_{kh} \equiv H^i_{ikh}, \quad \text{(b) } H_k \equiv H^i_{ik}, \text{ (c) } (n-1)\,H = H^i_i, \quad (9.20)$$[24]

we also get the following relations:

$$\left.\begin{array}{ll} \text{(a) } H_{kh}\, \dot{x}^h = H_k, & \text{(b) } H_{kh} = \dot{\partial}_h\, H_k, \\ \text{(c) } H_k\, \dot{x}^k = H^i_{ik}\, \dot{x}^k = H^i_i = (n-1)H. & \end{array}\right\} \quad (9.21)$$

Transvection of Eq. (9.21b) by $\dot{x}^k$ and application of Eq. (9.21c) also yields the relation

$$\dot{x}^k H_{kh} = (n-1)\dot{\partial}_h H - H_h. \quad (9.22)$$

The fundamental function F, the vector $\dot{x}^i$ and the unit vectors l^i and l_j given by Eqs. (2.5) and (2.6) remain invariant under Berwald's covariant differentiation. Hence, application of the commutation rule in Eq. (7.14) for F and use of Eq. (9.15b) also yields

$$l_i H^i_{jk} = 0. \quad (9.23)$$

The partial derivation of Eq. (9.18) w.r.t. $\dot{x}^k$ also yields a relation

$$\dot{x}^j\, \dot{\partial}_k\, H^i_j + H^i_k = 0, \quad (9.24)$$

which, on contraction w.r.t. i and k also gives

$$\dot{x}^j\, \dot{\partial}_i\, H^i_j + (n-1)\, H = 0, \quad (9.25)$$

by Eq. (9.21c). Contracting the indices i and k in Eq. (9.17), one can also derive

$$\dot{\partial}_r H^r_j - \dot{\partial}_j H^r_r = 3H^r_{jr} \Rightarrow \dot{\partial}_r H^r_j = (n-1)\,\dot{\partial}_j H - 3H_j, \qquad (9.26)$$

by Eqs. (9.20b) and (9.21c).

§ 10. Bianchi identities satisfied by various curvature tensors

A large number of curvature tensors discussed in the preceding sections satisfy identities analogous to Bianchi's first identity of Riemannian geometry:

$$\text{(a)} \qquad \Omega^i_{[j\,k\,h]} = 0, \qquad\qquad \text{(b)} \quad \Omega_{[j\,k\,h]l} = 0, \text{ [25]} \qquad (10.1)$$

where $\Omega^i_{j\,k\,h}$ is any curvature tensor other than $P^i_{j\,k\,h}$ and $R^i_{j\,k\,h}$. Such identities for $P^i_{j\,k\,h}$ are quite complicated. But, for the curvature tensor $R^i_{j\,k\,h}$, we may derive from Eq. (7.13), in consequence of Eqs. (9.8b) and (10.1a) used for the tensor $K^i_{j\,k\,h}$:

$$R^i_{[j\,k\,h]} = H^l_{[j\,k}\; C^i_{h]l}. \qquad (10.2)$$

Also, from Eq. (10.1a), applied for $H^i_{j\,k\,h}$, and contracted for index i and any lower index there follows, in consequence of Eq. (9.20a),

$$2H_{[k\,h]} = H^i_{h\,k\,i}. \qquad (10.3)$$

Thus, in contrast to Riemannian geometry, the tensor $H_{k\,h}$ is not, in general, symmetric in its indices.

In the following, we derive the identities analogous to Bianchi's second identity of Riemannian geometry. It is interesting to note that the curvature tensors $\widetilde{K}^i_{j\,k\,h}$ and $S^i_{j\,k\,h}$ satisfy identities exactly similar [26] to those in Riemannian geometry. On the other hand, the identities satisfied by the curvature tensors $K^i_{j\,k\,h}$, $H^i_{j\,k\,h}$ and $R^i_{j\,k\,h}$ are the generalizations of Bianchi's second identity. However, the tensor $H^i_{j\,k}$ again satisfies an identity similar to Bianchi's second identity of Riemannian geometry.

Theorem 10.1. The curvature tensors satisfy

$$\nabla_{[j}K^{i}_{kh]m} + (\dot{\partial}_l \Gamma^{*i}_{m[j}) H^{l}_{kh]} = 0, \tag{10.4}$$

$$\mathfrak{B}_{[j}H^{i}_{kh]m} + G^{i}_{lm[j} H^{l}_{kh]} = 0, \tag{10.5}$$

$$\mathfrak{R}_{[j}\widetilde{K}^{i}_{kh]m} = 0 \quad (10.6); \quad \dot{\nabla}_{[j}S^{i}_{kh]m} = 0. \tag{10.7}$$

Proof. Let Y_m be an arbitrary covariant vector field. Applying the commutation formula (7.11) to it and noting Eq. (9.8b), we have

$$2\nabla_{[k}\nabla_{h]}Y_m = -(K^{i}_{khm}Y_i + H^{i}_{kh}\dot{\partial}_i Y_m). \tag{10.8}$$

Operating it further by ∇_j, interchanging the indices j, k, h cyclically and rearranging the terms, the left member becomes

$$2\nabla_{[j}\nabla_{k]}(\nabla_h Y_m) + \text{cycl.}[j\,k\,h],$$

where cycl. $[j\,k\,h]$ denotes the sum of additional two terms obtained by the cyclic interchange of the indices j, k, h in the preceding terms. Applying the commutation formula (10.8) for the tensor fields $\nabla_h Y_m$, etc. and using the identity (10.1a) for the tensor K^{i}_{jkh}, above expression simplifies to

$$-\{K^{i}_{jkm}(\nabla_h Y_i) + (\dot{\partial}_i \nabla_h Y_m) H^{i}_{jk}\} + \text{cycl.}[j\,k\,h].$$

On the other hand, the right hand side of Eq. (10.8), under the same operation, yields

$$-\{Y_i(\nabla_j K^{i}_{khm}) + (\dot{\partial}_i Y_m)(\nabla_j H^{i}_{kh})\}$$
$$-\{K^{i}_{khm}\nabla_j Y_i + H^{i}_{kh}(\nabla_j \dot{\partial}_i Y_m)\} + \text{cycl.}[j\,k\,h].$$

Thus, cancelling certain common terms on either side, there follows

$$Y_i \nabla_j K^{i}_{khm} - (\dot{\partial}_i Y_m)\nabla_j H^{i}_{kh} + H^{i}_{kh}(\nabla_j \dot{\partial}_i - \dot{\partial}_i \nabla_j) Y_m + \text{cycl.}[j\,k\,h] = 0.$$

Applying the commutation rule vide Eq. (8.2) to the vector Y_m, and collecting the coefficients of like terms, it reduces to

$$Y_i\{\nabla_j K^{i}_{khm} + (\dot{\partial}_a \Gamma^{*i}_{jm}) H^{a}_{kh}\} + (\dot{\partial}_i Y_m)\{\nabla_j H^{i}_{kh} + (\nabla C^{i}_{aj}) H^{a}_{kh}\}$$
$$+ \text{cycl.}[j\,k\,h] = 0. \tag{10.9}$$

Putting $\dot{x}^m\, Y_m \equiv \phi$, so that $\dot{\partial}_i\, \phi = \dot{x}^m\, \dot{\partial}_i\, Y_m + Y_i$, a transvection of Eq. (10.9) with $\dot{x}^m$, for Eqs. (5.16), (6.4), (9.8b) and above relation, yields

$$(\dot{\partial}_i\, \phi)\,\{\nabla_j H^i_{kh} + (\nabla C^i_{aj}) H^a_{kh} + \text{cycl.}[j\,k\,h]\} = 0$$

$$\Rightarrow \qquad \{\nabla_j H^i_{kh} + (\nabla C^i_{aj}) H^a_{kh} + \text{cycl.}[j\,k\,h]\} = 0,$$

for being arbitrary. Hence, Eq. (10.9) reduces to

$$Y_i\,\{\nabla_j\, K^i_{khm} + (\dot{\partial}_a\, \Gamma^{*i}_{jm})\, H^a_{kh} + \text{cycl.}[j\,k\,h]\} = 0$$

$$\Rightarrow \qquad \nabla_j\, H^i_{kh} + (\nabla C^i_{aj})\, H^a_{kh} + \text{cycl.}[j\,k\,h] = 0, \qquad (10.10)$$

for Y_i being arbitrary. Its skew-symmetric part w.r.t. indices j, k and h, thus reads as

$$\nabla_{[j}\, H^i_{kh]} + (\nabla C^i_{a\,[j})\, H^a_{kh]} = 0. \qquad (10.11)$$

For Eq. (6.4), above identity assumes the alternate form given by Eq. (10.4).

Starting with the corresponding processes of covariant derivation and proceeding on similar lines we can also establish the remaining identities enunciated in the theorem. //

Theorem 10.2. The curvature tensor R^i_{jkh} satisfies

$$\nabla_{[j} R^i_{kh]m} + P^i_{l[j|m|} H^l_{kh]} = 0. \qquad (10.12)$$

Proof. From Eqs. (7.13), (10.4) and (10.11), there follows the identity

$$\nabla_{[j} R^i_{kh]m} + \{\dot{\partial}_l \Gamma^{*i}_{m[j} - \nabla_{[j} C^i_{|l\,m|} + C^i_{mp} \nabla C^p_{l[j}\} H^l_{kh]} = 0.$$

Finally, putting from Eq. (7.8a), this identity reduces to Eq. (10.12). //

Theorem 10.3. The tensors H^i_{jk} and H^i_j satisfy

$$\mathfrak{B}_{[j} H^i_{kh]} = 0, \qquad (10.13)$$

$$2\mathfrak{B}_{[j} H^i_{k]} + \mathfrak{B} H^i_{jk} = 0, \quad (\mathfrak{B} \equiv \dot{x}^h\, \mathfrak{B}_h). \qquad (10.14)$$

Proof. For homogeneous properties the tensor defined by Eqs. (9.12) satisfies

$$G^i_{l\,h\,j}\,\dot{x}^l = G^i_{l\,h\,j}\,\dot{x}^h = G^i_{l\,h\,j}\,\dot{x}^j = 0. \tag{10.15}$$

Consequently, the identity (10.5), on transvection by $\dot{x}^m$ yields (10.13). Further transvection of (10.13) by $\dot{x}^h$, for Eqs. (6.7c) and (9.16), establishes Eq. (10.14). //

§ 11. Special Finsler manifolds

There exist many special types of Finsler manifolds. For instance, in Section 6, it is seen that affinely connected manifolds form a special type of Finsler manifold. Like Riemannian geometry, imposing certain restrictions on the curvature tensors various special types of Finsler manifolds have been studied. Notably, we consider here (i) isotropic manifolds and (ii) manifolds of constant (scalar) curvature. Study of symmetric, and recurrent Finsler manifolds is postponed to Chapter 7.

11.1. Isotropic manifolds

The scalar Riemannian curvature $R(x^i, \dot{x}^i, \lambda^i)$ of F_n at a point P (x^i) along two directions $\dot{x}^i$ and λ^i through P is defined by

$$R(x^i, \dot{x}^i, \lambda^i) = K_{i\,j\,k\,h}\dot{x}^i\lambda^j\dot{x}^k\lambda^h / 2g_{i[j}\, g_{k]h}\dot{x}^i\lambda^j\dot{x}^k\lambda^h. \tag{11.1}$$ [27)]

A point is called *isotropic* if the Riemannian curvature defined above becomes independent of the choice of the direction λ^i. As such, the Riemannian curvature at an isotropic point may be simply written as $R(x^i, \dot{x}^i)$ or, in accordance with the Footnote 4, more briefly as R. In case every point of F_n is isotropic the manifold is called an *isotropic Finsler manifold.* In view of Eqs. (3.10.2), (2.4), (2.5), there follows a necessary relation

$$R(l_j l_h - g_{j\,h}) - K_{i\,j\,k\,h}\, l^i l^k = 0, \tag{11.2a}$$

from Eq. (11.1) in an isotropic manifold. Transvecting Eq. (11.2a) by $g^{j\,m}$ and applying Eqs. (2.1), (2.4) and (9.1b), above relation yields

$$R(l_j\, l^m - \delta^m_j) = K^m_{i\,j\,k}\, l^i l^k. \tag{11.2b}$$

Contracting it for the indices m and j, there results

$$(n-1)\,R = K^{j}_{jik}\, l^i l^k = K_{ik}\, l^i l^k\,, \tag{11.3}$$

by Eqs. (2.5), (9.10) and the skew-symmetry of the curvature tensor K^{j}_{ijk} in the first two lower indices. Thus, the Riemannian curvature of an isotropic Finsler manifold is determined by Eq. (11.3). It is seen that the Eqs. (11.2) with R given by (11.3) are also sufficient conditions for the manifold to be isotropic. Using Eq. (9.8b), an equivalent relation to Eqs. (11.2) is also obtained in the form [28)]

$$H^{i}_{jk} \equiv K^{i}_{jkh}\dot{x}^h = (2/3)F^2(\delta^{i}_{[j} - l^i l_{[j})\,\dot{\partial}_{k]}R + 2FR\delta^{i}_{[j} l_{k]}. \tag{11.4}$$

Lemma 11.1. The Riemannian (or sectional curvature) R is also related to Berwald's scalar curvature H given by Eq. (9.21) ([189], p. 147):

$$R = H / F^2. \tag{11.5}$$

Proof. Contraction of Eq. (11.4) with respect to the indices i and j, and transvection by $\dot{x}^k$, for the relations (9.21) and

$$l_k\, \dot{x}^k = F \tag{11.6}$$

derived from Eqs. (2.3) and (2.5) yields the result. //

Note 11.1. For a space of constant sectional curvature R, Eq. (11.4) reduces to

$$H^{i}_{jk} = 2FR\,\delta^{i}_{[j}\, l_{k]} = 2R\,\delta^{i}_{[j}\, y_{k]}, \tag{11.7}$$

for

$$F\, l_k = y_k \tag{11.8}$$

derived from Eq. (2.2). Directional derivation of Eq. (11.7) w.r.t. $\dot{x}^h$, for Eqs. (2.7a) and (9.15a) also gives

$$H^{i}_{jkh} = 2R\,\delta^{i}_{[j}\, g_{k]h}. \tag{11.9}$$

Its contraction w.r.t. the indices i and j, for Eq. (9.20a), also yields

$$H_{kh} = R\, g_{kh}. \tag{11.10}$$

Foot-notes:

3) Indeed, $\dot{x}^i$'s span a tangent manifold T_n (P) at the point P (x^i). Cf. [189], p. 10.

4) Now onward unless stated otherwise all the entities will be supposed functions of the line-elements ($x^i, \dot{x}^i$); but, for simplicity, they will be written without assigning the same.

5) More precisely we mean the vector with components $\dot{x}^i$'s.

6) For instance, the tangent manifold T_n (P) possesses this property. Cf. [189], p. 11. For Minkowskian manifolds see [29].

7) Round (respectively square) brackets are used to denote the symmetric (resp. skew-symmetric) parts with regard to the indices enclosed therein and the indices within two solidi, if any, remain unaffected by the symmetry or skew-symmetry.

$$T_{(jk)} \equiv (1/2!)\,(T_{jk} + T_{jk}); \qquad T_{[jk]} \equiv (1/2!)\,(T_{jk} - T_{jk}).$$

8) By an object we mean a vector or a tensor throughout this section.

9) Analogous to the processes of covariant derivation given by Cartan and by Berwald (cf. Eqs. (5.14) and (6.6) in the following sections), the author ([104], Eq. (2.3)) defines the covariant derivation of an object for the generalized Christoffel symbols.

10) The directional arguments $\dot{x}^j$ of the line-elements $(x^j, \dot{x}^j)$ are here replaced by an arbitrary vector $\xi^j(x^k)$ of the tangent manifold T_n (P) with centre at P (x^k), cf. [189], pp. 10, 60.

11) The original notation for this covariant derivative given by Rund ([189], Eq. (2.5.1)) is $X^i_{;k}$ which, for convenience, is taken in the present form by the author [107].

12) For *ab initio* development of the theory of Cartan, see [189], pp. 65 - 71.

13) Cartan uses the terminology "element of support" for the line-element $(x^i, \dot{x}^i)$. But, Rund [189] makes it confusing by using the same terminology for both the line-element $(x^i, \dot{x}^i)$ (cf. p. 67) as well as its directional argument $\dot{x}^i$ (cf. p. 75).

14) The symbolism used for these covariant derivatives is due to Misra and Pande [142], and Yano ([234], p. 179), whereas the original notations used for them by Rund ([189], pp. 73, 99) and Cartan ([33], p. 34) are $(1/F)\, X^i\big|_k$ and $X^i_{\,|k}$ respectively. Matsumoto ([73], p. 87) calls the derivative defined by Eq. (5.13) the "*v-covariant derivative*" against the terminology "*partial derivative*" used by Rund (op. cit.) for the process illustrated by $X^i\big|_k = F\,\dot{\partial}_k X^i + X^j A^i_{jk}$.

15) (x^k, ξ^k) and $(x^k, \dot{x}^k)$ for Eqs. (4.7) and (5.14) respectively together with the coincidence of the vector fields ξ^k and $\dot{x}^k$ at the point P (x^k), where the covariant derivatives are evaluated.

16) Differentiating Eq. (6.2b) partially with respect to $\dot{x}^j$ and using Eq. (6.1) we easily derive the first of Eqs. (6.4). For the proof of the second one see [189], p. 81.

17) The notation for the covariant derivative, used here, is due to Misra and Pande [143] replacing the original notation $X^i_{(k)}$ of [189], p. 80.

18) The symbols $\widetilde{K}^i_{jkh}$ used here coincide with $\widetilde{K}^i_{hkj}$ of [189], Eq. (4.1.5).

19) This term is wrongly taken zero in [189], Eq. (4.1.23). In fact, the expression in the second member of the equation should be augmented by $2F^2\dot{\partial}_{[k}C^i_{h]r} = -4A^i_{m[k}A^m_{h]r}$. Consequently, in Eq. (4.1.24) of [189], either the sign in the last term be taken negative else the indices k and h should be interchanged in order to be in accordance with Eq. (4.1.25) of [189]. Similar corrections should also be made in [110], Eq. (1.8) and [142], Eq. (1.11). With the latter choice comparison of our equations (7.5) and (7.6) with (4.1.25) and (4.1.24) of [189] shows that the tensor S^i_{jkh} used here coincides with $(1/F^2)$. S^i_{hkj} of [189] and S^i_{hkj} of [73], p. 87. The expression in [73] is correct. However, it could have been reduced in aid of our Eq. (7.4). See also [203], last relation of Eq. (10.10).

20) As in [142] our P^i_{jkh} equals $(1/F)\, P^i_{hkj}$ of [189], Eq. (4.1.29).

[21)] Like [142], the tensor K^{i}_{jkh} defined here and the tensors R^{i}_{jkh}, H^{i}_{jkh} (to be introduced) coincide with the respective tensors K^{i}_{hkj}, R^{i}_{hkj} and H^{i}_{hkj} given by Eqs. (4.1.7), (4.1.34) and (4.6.7) of [189].

[22)] See Misra [100], [101].

[23)] The associate curvature tensors are defined here in different way to those in [189], p. 105. Our Ω_{jkhl} coincide with Ω_{hlkj} of [189] for the curvature tensors other than S^{i}_{jkh} and P^{i}_{jkh}. In case of the latter, S_{jkhl} and P_{jkhl}, used here, coincide with $(1/F^2)\, S_{hlkj}$ and $(1/F)\, P_{hlkj}$ respectively of [189].

[24)] As in [108], [109], the symbols H^{i}_{jk} and H_{kh}, used here, coincide with H^{i}_{kj} and H_{hk} respectively of [189], pp. 125, and 129. However, H^{i}_{j} and H_{k} used here are the same as in [189].

[25)] The symmetric (resp. skew-symmetric) parts of a geometric object Ω^{i}_{jkh} with respect to the indices *j*, *k*, *h* are denoted by

$$\Omega^{i}_{(jkh)} \equiv (1/3!)\,\{\Omega^{i}_{jkh} + \Omega^{i}_{khj} + \Omega^{i}_{hjk} + \Omega^{i}_{kjh} + \Omega^{i}_{hkj} + \Omega^{i}_{jhk}\},$$

(resp.)

$$\Omega^{i}_{[jkh]} \equiv (1/3!)\,\{\Omega^{i}_{jkh} + \Omega^{i}_{khj} + \Omega^{i}_{hjk} - \Omega^{i}_{kjh} - \Omega^{i}_{hkj} - \Omega^{i}_{jhk}\}.$$

[26)] Like Riemannian geometry, the skew-symmetric parts of the covariant derivatives of curvature tensors with respect to the lower indices vanish.

[27)] Cf. [189], Eq. (4.7.1). Rewriting this equation as per our notation explained in the Foot-note 21, canceling a negative sign from both numerator and denominator, and interchanging the dummy indices *k*, *j*, *h*, *i* cyclically, we get Eq. (11.1.).

[28)] Cf. [189], Eq. (4.7.6) and [112], Eq. (2.7). However, in [108] the order of the indices *j*, *k* on the right hand side of Eq. (2.6) should be interchanged.

CHAPTER 5

TRANSFORMATIONS IN FINSLER SPACE

§ 1. Paths and projective connection in a Finsler space

Analogous to Riemannian geometry [41], [92], [123], [231], the integrals of the second order differential equations (4.3.3) represent geodesics in a Finsler space, where s measures the arc-length and $\dot{x}^i$ form the components of tangent vector of the curve [189]. In the non-metrical approach of Berwald the curves analogous to geodesics have been called *paths*. In view of Eq. (4.5.3), the Eqs. (4.3.3) may be altered as

$$\ddot{x}^i + 2G^i = (\ddot{s}/\dot{s})\dot{x}^i, \tag{1.1}$$

representing the differential equations of paths. Multiplying it by $\dot{x}^h$ and forming its skew-symmetric part w.r.t. the indices i and h, the factor $\ddot{s}/\dot{s}$ gets eliminated:

$$\ddot{x}^i\,\dot{x}^h + 2G^i\,\dot{x}^h = \ddot{x}^h\,\dot{x}^i + 2G^h\,\dot{x}^i. \tag{1.2}$$

This is the exact generalization of Eq. (15.27.2) of [123]. Now subjecting the space F_n under some transformation, the differential equations of paths of the deformed space $\overline{F}_n$ can be similarly written:

$$\ddot{x}^i\,\dot{x}^h + 2\overline{G}^i\,\dot{x}^h = \ddot{x}^h\,\dot{x}^i + 2\overline{G}^h\,\dot{x}^i. \tag{1.3}$$

The transformations $F_n \to \overline{F}_n$ will map the paths of F_n onto the paths of $\overline{F}_n$ if the differential equations (1.2) and (1.3) remain same. For this, we must have

$$(\overline{G}^i - G^i)\,\dot{x}^h - (\overline{G}^h - G^h)\,\dot{x}^i = 0. \tag{1.4}$$

Analogous to [123] it may be seen that above equations possess a solution

$$\overline{G}^i = G^i + p(x, \dot{x})\,\dot{x}^i, \tag{1.5}$$

where p are some arbitrary scalar functions of the line-elements positively homogeneous of degree 1 in their directioanl arguments. Following the terminology used in the restricted geometry of paths, i.e. the non-Riemannian geometry, we may call above change of functions G^i as the *projective change*. The partial derivatives of Eq. (1.5) w.r.t. the directional coordiantes also yield

$$\overline{G}^i_j = G^i_j + p\,\delta^i_j + p_j\,\dot{x}^i, \tag{1.6}$$

and

$$\overline{G}^i_{jk} = G^i_{jk} + 2\,\delta^i_{(j}\,p_{k)} + p_{jk}\,\dot{x}^i, \tag{1.7}$$

where the entities

$$2\,\delta^i_{(j}\,p_{k)} \equiv \delta^i_j\,p_k + p_j\,\delta^i_k, \tag{1.8}$$

$$p_j \equiv \dot{\partial}_j\,p \quad (1.9a); \qquad p_{jk} \equiv \dot{\partial}_k\,\dot{\partial}_j\,p, \quad (1.9b)$$

form components of a tensor, vector and tensor respectively. They are positively homogeneous functions of degrees 0, 0 and – 1 respectively in $\dot{x}^i$'s. Hence, they satisfy equations

$$p_i\,\dot{x}^i = p \quad (1.10a); \qquad p_{ji}\,\dot{x}^i = 0. \quad (1.10b)$$

1.1. Projective connection parameters

Contracting Eq. (1.7) w.r.t. indices *i* and *k*, and using Eq. (1.10b), we get

$$\overline{G}^i_{ji} - G^i_{ji} = (n+1)\,p_j. \tag{1.11}$$

Its directional derivation w.r.t. $\dot{x}^k$, for Eq. (4.9.12) and (1.9b) yields

$$\overline{G}^i_{jik} - G^i_{jik} = (n+1)\,p_{jk}. \tag{1.12}$$

Elimination of derivatives p_j and p_{jk} from Eq. (1.7) by means of Eqs. (1.11) and (1.12) gives

$$\overline{G}^i_{jk} - G^i_{jk} = [\,2\delta^i_{(j}\,\{\overline{G}^r_{k)r} - G^r_{k)r}\} + \dot{x}^i\,\{\overline{G}^r_{jkr} - G^r_{jkr}\}\,]/(n+1),$$

i.e.

$$\Pi^i_{jk} \equiv \overline{G}^i_{jk} - \{2\delta^i_{(j}\,\overline{G}^r_{k)r} + \dot{x}^i\,\overline{G}^r_{jkr}\}/(n+1)$$

$$= G^i_{jk} - \{2\delta^i_{(j}\,G^r_{k)r} + \dot{x}^i\,G^r_{jkr}\}/(n+1) \tag{1.13}$$

remains invariant under the projective change given by Eq. (1.5).

Definition 1.1. The functions defined by Eq. (1.13) are called the *projective connection parameters*.

The tensor $G^{r}_{j\,k\,r}$ being symmetric in its lower indices the projective connection parameters are also symmetric in *j*, *k* and are positively homogeneous of degree zero in their directional arguments. Certain properties of projective connection parameters are derived by the author et al. [131]. It is seen there [131], Eq. (2.2) that the contracted symbols $\Pi^{i}_{j\,i}$ vanish identically and their transvected part with $\dot{x}^k$ is

$$\Pi^{i}_{j} \equiv \Pi^{i}_{j\,k}\, \dot{x}^k = G^{i}_{j} - (\delta^{i}_{j}\, G^{r}_{r} + \dot{x}^{i}\, G^{r}_{r\,j}) / (n+1), \quad (1.14)$$

by Eqs. (4.6.2a) and (4.10.15). However, their directional derivative:

$$\Pi^{i}_{j\,k\,h} \equiv \dot{\partial}_h\, \Pi^{i}_{j\,k} = G^{i}_{j\,k\,h} - \{3\delta^{i}_{(j}\, G^{r}_{k\,h)\,r} + \dot{x}^{i}\, G^{r}_{j\,k\,r\,h}\}/(n+1), \quad (1.15)$$

is a tensor, symmetric in its all lower indices. For homogeneous properties, it satisfies

$$\Pi^{i}_{j\,k\,h}\, \dot{x}^h = 0. \quad (1.16)$$

The contracted part of this tensor also vanished identically [131], Eq. (2.6):

$$\Pi^{i}_{i\,k\,h} = \Pi^{i}_{j\,i\,h} = \Pi^{i}_{j\,k\,i} = 0. \quad (1.17)$$

It is observed by the author et al. [131] that the coordinate transformation laws for these connection parameters are not similar to those for the Berwald's connection parameters $G^{i}_{j\,k}$. Instead their coordinate transformation laws involve two additional terms [131], Eq. (3.1):

$$\overline{\Pi}^{i}_{j\,k} = (\partial_a\, \bar{x}^{i}) \left\{ \Pi^{a}_{b\,c} (\bar{\partial}_j\, x^b)(\bar{\partial}_k\, x^c) + \bar{\partial}_j\, \bar{\partial}_k\, x^a \right\}$$
$$+ (\bar{\partial}_r\, x^a)(\partial_a\, \partial_b\, \bar{x}^r) \left\{ \delta^{i}_{j}\, (\bar{\partial}_k\, x^b) + \delta^{i}_{k}\, (\bar{\partial}_j\, x^b) \right\} / (n+1), \quad (1.18)$$

where

$$\partial_a \equiv \partial / \partial x^a, \qquad \bar{\partial}_j \equiv \partial / \partial \bar{x}^j. \quad (1.19)$$

Analogous to Berwald's covariant derivation process defined by Eq. (4.6.6), the projective covariant derivative of vector field with respect to positional coordinates has been defined by the author [98], Eq. (2.2):

$$X^i_{((k))} \equiv \partial_k X^i - (\dot{\partial}_j X^i) \Pi^j_{kh} \dot{x}^h + X^j \Pi^i_{jk}. \quad (1.20)$$

The alternate notation for this derivative was also used as $\mathcal{P}_k X^i$ by the author et al. [131], p. 59. It is also seen there that the derivative vanishes identically for the vector field $\dot{x}^i$:

$$\mathcal{P}_k \dot{x}^i = 0, \quad (1.21)$$

as is the case with Berwald's covariant derivation (cf. Eq. (4.6.7)). However, unlike to Berwald's case, the projective covariant derivatives of the metric function F and the unit vector l^i along $\dot{x}^i$ do not vanish [98], Eqs. (2.7), (2.8):

$$\mathcal{P}_k l^i = -l^i \left(l_k G^r_r + F G^r_{rk}\right)/(n+1) F, \quad (1.22a)$$

and

$$\mathcal{P}_k F = \left(l_k G^r_r + F G^r_{rk}\right)/(n+1). \quad (1.22b)$$

It may be noted that such covariant derivation process does not preserve the tensorial character of the objects. Onward theory of commutation rules and subsequent curvature tensor like quantities (denoted by Q^i_{hkj} (or in view of alternate notation for the covariant derivative, by Q^i_{jkh}) are also developed therein which are not tensors:

$$2\,\mathcal{P}_{[j} \mathcal{P}_{k]} X^i = Q^i_{jkh} X^h - (\dot{\partial}_h X^i) Q^h_{jkm} \dot{x}^m, \quad (1.23)$$

where

$$Q^i_{jkh} = 2\{\partial_{[j} \Pi^i_{k]h} + (\dot{\partial}_l \Pi^i_{h[j}) \Pi^l_{k]} + \Pi^i_{l[j}) \Pi^l_{k]h}\}, \quad (1.24)$$

where the functions Π^l_k are given by Eq. (1.14). Transvections and contractions of these projective quantities also gave rise to the quantities:

$$Q^i_{jk} \equiv Q^i_{jkh} \dot{x}^h = 2\{\partial_{[j} \Pi^i_{k]} + \Pi^i_{l[j}) \Pi^l_{k]}\}, \quad (1.25)$$

and

$$Q^i_j \equiv Q^i_{jk} \dot{x}^k; \qquad Q_{kh} \equiv Q^i_{ikh}, \quad Q_k \equiv Q^i_{ik}. \quad (1.26)$$

Note 12.1. Dropping the first two symmetric terms involving the Kronecker deltas in Eq. (1.13), Yano [234], p. 196, defined a new set of parameters, called *normal projective connection* parameters (cf. Eq. (10.4.1) below), the corresponding covariant derivation process, commutation rules and the curvature tensor, called the *normal projective curvature tensor* $N^{i}_{j\,k\,h}$ have been subsequently developed by him.

§ 2. Projective curvature tensor

Constructing a tensor $\overline{H}^{i}_{j}$ of $\overline{F}_{n}$ in analogy with H^{i}_{j} defined by Eq. (4.9.16):

$$\overline{H}^{i}_{j} \equiv \overline{H}^{i}_{j\,k}\,\dot{x}^{k} = 2\partial_{j}\,\overline{G}^{i} - \dot{x}^{k}\partial_{k}\,\overline{G}^{i}_{j} + 2\overline{G}^{i}_{l\,j}\,\overline{G}^{l} - \overline{G}^{i}_{l}\,\overline{G}^{l}_{j}\,. \quad (2.1)$$

Putting for the values of transformed functions from Eqs. (1.5) - (1.7), using Eq. (4.9.16) and homogeneous properties of functions, above equation simplifies to

$$\overline{H}^{i}_{j} = H^{i}_{j} + \dot{x}^{i}\,\{2\partial_{j}\,p - \dot{x}^{k}\partial_{k}\,p_{j} + 2\,p_{l\,j}\,G^{l} - p_{l}\,G^{l}_{j} - p\,p_{j}\}$$

$$+\ \delta^{i}_{j}\,\{-\dot{x}^{k}\,\partial_{k}\,p + 2\,p_{l}\,G^{l} + p^{2}\}$$

$$= H^{i}_{j} + \dot{x}^{i}\,\{2\mathfrak{B}_{j}\,p\ - \dot{x}^{k}\,\mathfrak{B}_{k}\,p_{j}\ - p\,p_{j}\} + \delta^{i}_{j}\,\{-\dot{x}^{k}\,\mathfrak{B}_{k}\,p\ + p^{2}\}, \quad (2.2)$$

by covariant derivation formula (4.6.6) applied to scalar function p and its derivative p_j given by Eq. (1.9). Contraction of Eq. (2.2) with respect to indices i and j, for Eq. (4.9.21c), and division by $n-1$ also determines

$$\overline{H} = H\ - \dot{x}^{k}\,\mathfrak{B}_{k}\,p\ +\ p^{2}. \quad (2.3)$$

Subtraction of δ^{i}_{j} multiple of Eq. (2.3) from Eq. (2.2) gives

$$\overline{H}^{i}_{j}\ - \overline{H}\ \delta^{i}_{j} = H^{i}_{j} - H\,\delta^{i}_{j} + \dot{x}^{i}\,U_{j}\,, \quad (2.4)$$

where, for convenience, we have put

$$U_{j} \equiv 2\mathfrak{B}_{j}\,p\ - \dot{x}^{k}\,\mathfrak{B}_{k}\,p_{j}\ - p\,p_{j}\}. \quad (2.5)$$

The partial derivation of Eq. (2.4) w.r.t. $\dot{x}^{h}$ and then contraction of indices i and h, for homogeneous properties of the vector U_j, yields

$$\dot{\partial}_{r}\,\overline{H}^{r}_{j} - \dot{\partial}_{j}\,\overline{H} = \dot{\partial}_{r}\,H^{r}_{j} - \dot{\partial}_{j}\,H + (n+1)\,U_{j}\,. \quad (2.6)$$

Finally, elimination of the term containing U_j from Eqs. (2.4) and (2.6), gives a tensor, which is invariant under the projective change:

$$W^i_j \equiv \overline{H}^i_j - \overline{H}\,\delta^i_j - \dot{x}^i(\dot{\partial}_r \overline{H}^r_j - \dot{\partial}_j \overline{H})/(n+1)$$

$$= H^i_j - H\,\delta^i_j - \dot{x}^i(\dot{\partial}_r H^r_j - \dot{\partial}_j H)/(n+1). \qquad (2.7)$$

Definition 2.1. The tensor defined by Eq. (2.7) is positively homogenous of degree 2 in $\dot{x}^i$'s and is called *projective deviation tensor* in analogy with the Berwald's deviation tensor H^i_j defined by Eq. (4.9.16).

2.1. Properties of W^i_j

Analogous to Berwald's deviation tensor satisfying Eq. (4.9.18) the projective deviation tensor also vanishes on transvection by $\dot{x}^j$:

$$W^i_j\,\dot{x}^j = 0. \qquad (2.8)$$

On contrary to Eq. (4.9.21c), the contracted part of W^i_j w.r.t. indices i and j vanishes identically:

$$W^i_i = (n-1)\,H - nH + (n-1)\,H/(n+1) + 2\,H/(n+1) = 0. \qquad (2.9)$$

Further, derivation of Eq. (2.8) w.r.t. $\dot{x}^k$ yields

$$\dot{x}^j\,\dot{\partial}_k W^i_j + W^i_k = 0, \qquad (2.10)$$

in analogy with Eq. (4.9.24). On the other hand, derivation of Eq. (2.7) w.r.t. $\dot{x}^k$ gives

$$\dot{\partial}_k W^i_j = \dot{\partial}_k H^i_j - \delta^i_j\,\dot{\partial}_k H - \delta^i_k(\dot{\partial}_r H^r_j - \dot{\partial}_j H)/(n+1)$$

$$-\dot{x}^i\,\dot{\partial}_k(\dot{\partial}_r H^r_j - \dot{\partial}_j H)/(n+1)\,. \qquad (2.11)$$

Its contraction w.r.t. the indices i and k, for the homogeneous properties of the vector within parentheses determines

$$\dot{\partial}_i W^i_j = \dot{\partial}_i H^i_j - \dot{\partial}_j H - (n+1)\,(\dot{\partial}_i H^i_j - \dot{\partial}_j H)/(n+1) = 0. \ (2.12)$$

2.2. Tensor W^i_{jk}

Now we are in position to introduce the other projective tensors associated to W^i_j in analogy with Berwald's curvature tensors discussed in previous chapter. Thus, in analogy with Eq. (4.9.17), we first construct the tensor

$$W^i_{jk} = (2/3)\dot{\partial}_{[k} W^i_{j]}, \tag{2.13a}$$

which is positively homogenous of degree 1 in $\dot{x}^i$'s. Thus, taking skew-symmetric part of the tensors in Eq. (2.11) and putting from Eqs. (4.9.17) and (4.9.26),

$$\begin{aligned} W^i_{jk} &= H^i_{jk} + \{\dot{x}^i \dot{\partial}_r H^r_{kj} - \delta^i_j(\dot{\partial}_k H + H_k) + \delta^i_k(\dot{\partial}_j H + H_j)\}/(n+1) \\ &= H^i_{jk} - (\dot{x}^i H^r_{jkr})/(n+1) \\ &+ \{\delta^i_k(\dot{x}^r H_{rj} + nH_j) - \delta^i_j(\dot{x}^r H_{rk} + nH_k)\}/(n^2-1), \end{aligned} \tag{2.13b}$$

by Eqs. (4.9.15a), (4.9.22) and (4.10.3). Employing the skew-symmetric notation of Foot-note 7, it can be written more compactly as

$$\begin{aligned} W^i_{jk} &= H^i_{jk} - (\dot{x}^i H^r_{jkr})/(n+1) \\ &+ 2\{n H_{[j} + \dot{x}^r H_{r[j}\}\delta^i_{k]} / (n^2-1). \end{aligned} \tag{2.14}$$

Finally, in analogy with Eq. (4.9.15a), we construct the projective curvature tensor as follows:

$$W^i_{jkh} \equiv \dot{\partial}_h W^i_{jk} = (2/3)\, \dot{\partial}_h \dot{\partial}_{[k} W^i_{j]}. \tag{2.15}$$

Differentiating Eq. (2.14) w.r.t. $\dot{x}^h$ and putting from Eqs. (4.9.15a), we derive

$$\begin{aligned} W^i_{jkh} &= H^i_{jkh} - (H^r_{jkr}\delta^i_h + \dot{x}^i \dot{\partial}_h H^r_{jkr})/(n+1) \\ &+ 2(n\dot{\partial}_h H_{[j} + H_{h[j} + \dot{x}^r \dot{\partial}_h H_{r[j})\delta^i_{k]} / (n^2-1), \end{aligned} \tag{2.16}$$

2.3. Properties of projective tensors

The tensor defined by Eq. (2.15) is also positively homogenous of degree zero in $\dot{x}^i$'s. Hence, in analogy with Eqs. (4.9.15b) and (4.9.16), the projective tensors are also related to each other:

$$W^i_{jkh}\,\dot{x}^h = W^i_{jk} \qquad (2.17); \qquad W^i_{jk}\,\dot{x}^k = W^i_j\,, \qquad (2.18)$$

where the Eqs. (2.8) and (2.10) have been used to establish the second relation (2.18).

Contracting the tensor W^i_{jk} defined by Eq. (2.14) w.r.t. indices i and k and putting from Eqs. (2.9) and (2.12), we also get

$$W^i_{ji} = 0, \qquad (2.19)$$

which, on derivation w.r.t. $\dot{x}^h$ also yields

$$W^i_{jih} = 0. \qquad (2.20)$$

Further, contraction of indices i and h in Eq. (2.15), for Eq. (2.12), also makes the contracted tensor zero:

$$W^i_{jki} = 0. \qquad (2.21)$$

For skew-symmetric properties of the tensors W^i_{jk} and W^i_{jkh} in indices j, k Eqs. (2.19) - (2.21) can be compounded together:

$$W^i_{ji} = -W^i_{ij} = W^i_{jih} = -W^i_{ijh} = W^i_{jki} = 0. \qquad (2.22)$$

Thus, all the contracted parts of these tensors (irrespective of any lower index and the upper index) vanish identically.

Theorem 2.1. The projective deviation tensor vanishes identically in an isotropic Finsler space.

Proof. Transvection of Eq. (4.11.4) with the vector $\dot{x}^k$, for Eqs. (4.2.3), (4.9.16) and the homogeneous properties of the Riemannian (scalar) curvature R in its directional arguments yields after simplification

$$H^i_j = F^2 R(\delta^i_j - l^i\, l_j) = H\,(\delta^i_j - l^i\, l_j), \qquad (2.23)$$

by Eq. (4.11.5). Its derivation w.r.t. $\dot{x}^k$:

$$\dot{\partial}_k H^i_j = \dot{\partial}_k H\,(\delta^i_j - l^i\, l_j) - H\{(\dot{\partial}_k l^i)\, l_j + l^i (\dot{\partial}_k l_j)\},$$

on contraction w.r.t. indices i and k, for Eqs. (4.2.5), (4.2.8), and homogeneous properties of H and l_j yields

$$\dot{\partial}_r H^r_j = \dot{\partial}_j H - (2H/F)l_j - (n-1)(H/F)l_j$$

$$\Rightarrow \qquad \dot{\partial}_r H^r_j - \dot{\partial}_j H = -(n+1)(H/F)l_j. \tag{2.24}$$

Substitution from Eqs. (4.11.6), (2.23) and (2.24) in Eq. (2.7) yields the projective deviation tensor

$$W^i_j = H(\delta^i_j - l^i l_j) - H\delta^i_j + \dot{x}^i (H/F) l_j = 0. \text{ //} \tag{2.25}$$

Corollary 2.1. The projective tensors W^i_{jk} and W^i_{jkh} also vanish identically in an isotropic Finsler space.

$$W^i_{jk} = W^i_{jkh} = 0. \tag{2.26}$$

Proof. The results follow immediately from the relations (2.15). //

Note 2.1. It may be easily verified from Eqs. (2.18) and (2.22) that the tensors W^i_j, W^i_{jk} and W^i_{jkh} vanish identically in a 2-dimensional Finsler space.

The author also derived a relation connecting the projective curvature tensor W^i_{jkh} with projective quantities Q^i_{jkh} [131], Eq. (2.14):

$$W^i_{jkh} = Q^i_{jkh} + \{Q_{hj}\,\delta^i_k - Q_{hk}\,\delta^i_j\}/(n-1). \tag{2.27}$$

§ 3. Conformal transformation

Conformal transformations in Finslerian spaces are studied by Hiramatu [48], [49], Izumi [52], [53], Pandey [171], [173], Pandey and Dwivedi [182], present author [99], [102], [106], present author with Fava [126] and others [189]. Study of conformal transformations in other special types of Finslerian spaces is scarcely explored.

Definition 3.1. The transformations of the space preserving angle between two directions are called *conformal*.

It may be noted that these transformations do not affect the positional as well as the directional coordinates. Let $F(x, \dot{x})$ and $\overline{F}(x, \dot{x})$ be the metric functions of two spaces F_n and $\overline{F}_n$ satisfying all the conditions in § 7 of Chapter 3, giving rise to the corresponding metric tensors g_{ij} and $\overline{g}_{ij}$. The correspondence $\mathfrak{T}$: $F_n \to \overline{F}_n$ will be conformal if the angle φ between two arbitrary directions, say X^i and Y^i remains same in both the spaces. But,

$$\cos\varphi = g_{ij}\, X^i\, Y^j \,/\, \sqrt{g_{ab}\, X^a\, X^b}\,.\sqrt{g_{cd}\, Y^c\, Y^d}, \quad (3.1)$$

remains invariant if the metric tensors of two spaces are proportional:

$$\overline{g}_{ij} = \psi(x, \dot{x})\, g_{ij}. \tag{3.2a}$$

Derivation of this equation w.r.t. $\dot{x}^k$, for Eq. (3.9.1) and its analogue in $\overline{F}_n$ gives

$$2\overline{C}_{ijk} = 2\psi(x, \dot{x})\, C_{ijk} + (\dot{\partial}_k\, \psi)\, g_{ij}. \tag{3.3}$$

Noting the symmetry of tensors C_{ijk} and $\overline{C}_{ijk}$, the skew-symmetric part of above equation w.r.t. indices i and k yields

$$(\dot{\partial}_k\, \psi)\, g_{ij} + (\dot{\partial}_i\, \psi)\, g_{kj} = 0.$$

Transvecting it by g^{jk} and using Eqs. (4.2.1), there follows:

$$(n+1)(\dot{\partial}_i\, \psi) = 0 \quad \Rightarrow \quad (\dot{\partial}_i\, \psi) = 0. \tag{3.4}$$

Thus, the factor of proportionality in Eq. (3.2a) is at most a point function. For convenience, we take it as an exponential function

$$\psi = e^{2\sigma} \quad \Rightarrow \quad \sigma = \ln\psi. \tag{3.5}$$

Hence, Eq. (3.2a) reduces to

$$\overline{g}_{ij} = e^{2\sigma}\, g_{ij}. \tag{3.2b}$$

As a result, the associate metric tensors of the two spaces are connected by

$$\overline{g}^{ij} = e^{-2\sigma}\, g^{ij}. \tag{3.6}$$

Also, we get

$$\overline{F} = e^{\sigma}\, F. \tag{3.7}$$

Further, in view of Eqs. (3.4) and (3.5), the Eq. (3.3) reduces to

$$\overline{C}_{ijk} = e^{2\sigma}\, C_{ijk}, \qquad (3.8)$$

determining the conformal change in the tensor C_{ijk}. From Eqs. (3.7) and (3.8), there also results

$$\overline{A}_{ijk} = e^{3\sigma}\, C_{ijk}, \qquad (3.9)$$

Transvection of Eqs. (3.8) and (3.9) by the tensor $\overline{g}^{hj}$, for Eqs. (3.6), (4.2.11) and (4.2.12) establishes:

$$\overline{C}^{h}_{ik} = C^{h}_{ik} \qquad (3.10); \quad \text{and} \quad \overline{A}^{h}_{ik} = e^{\sigma} A^{h}_{ik}. \quad (3.11)$$

Theorem 3.1. The tensor C^{i}_{jk} remains invariant under the conformal change.

Note 3.1. Some more quantities remaining invariant under conformal change are evaluated by the author et al. [126]. Notably, the following result has been proved therein:

Theorem 3.2. The second kind Eulerian curvature tensor [29)] is a conformal invariant.

3.1. Conformal change in connection parameters

The generalized Christoffel symbols given by Eqs. (4.3.1) transform under the conformal change according to

$$\overline{\gamma}_{ikj} = \partial_{(i}\, \overline{g}_{j)k} - (1/2)\, \partial_{k}\, \overline{g}_{ij} = e^{2\sigma}\{\gamma_{ikj} + 2\sigma_{(i}\, g_{j)k} - \sigma_{k}\, g_{ij}\}, \quad (3.12)$$

by Eqs. (3.2b), (3.4); and, for convenience, we have written

$$\sigma_{k} \equiv \partial_{k}\, \sigma. \qquad (3.13a)$$

This is a gradient vector field independent of $\dot{x}^{i}$'s. Further, the second kind symbols defined by Eq. (4.3.2), for Eqs. (3.4), (3.6) and

$$\sigma^{h} \equiv g^{kh}\, \sigma_{k}, \qquad (3.13b)$$

transform to

$$\overline{\gamma}^{h}_{ij} = \gamma^{h}_{ij} + 2\sigma_{(i}\, \delta^{h}_{j)} - \sigma^{h}\, g_{ij}. \qquad (3.14)$$

Considering the second order directional derivation of the functions given by Eq. (4.5.3), using symmetry of the Christoffel symbols in their lower indices and simplifying by means of Eq. (4.5.3) itself, we derive

$$G^{i}_{jk} = \gamma^{i}_{jk} + 2\,\dot{x}^{l}\,\dot{\partial}_{(j}\,\gamma^{i}_{k)l} + (1/2)\dot{x}^{l}\,\dot{x}^{m}\,\dot{\partial}_{j}\,\dot{\partial}_{k}\,\gamma^{i}_{lm}. \quad (3.15)$$

Using Eqs. (3.14), (4.2.9) and independence of the vector σ_k on directional arguments above functions transform according to

$$\overline{G}^{i}_{jk} = G^{i}_{jk} + 2\delta^{i}_{(j}\,\sigma_{k)} - \sigma^{i} g_{jk} - 2y_{(j}\,\dot{\partial}_{k)}\sigma^{i} - (F^{2}/2)\,\dot{\partial}_{j}\dot{\partial}_{k}\,\sigma^{i}. (3.16)$$

Introducing a tensor

$$B^{im} = (F^{2}/2)\,g^{im} - \dot{x}^{i}\,\dot{x}^{m}\,, \quad (3.17)$$

and computing its second order derivation w.r.t. directional coordinates we may also derive

$$\dot{\partial}_{j}\dot{\partial}_{k}B^{im} = (\dot{\partial}_{j}y_{k})\,g^{im} + 2y_{(j}\,\dot{\partial}_{k)}\,g^{im} + (F^{2}/2)\dot{\partial}_{j}\dot{\partial}_{k}g^{im} - 2\delta^{i}_{(j}\delta^{m}_{k)}\,.$$

On transvection with σ_m, using Eq. (3.13b), (4.2.7a) and independence of σ_m it can be verified that the expression for the tensor $-(\dot{\partial}_{j}\dot{\partial}_{k}B^{im})\,\sigma_{m}$ coincides with the terms giving variation of G^{i}_{jk} given by Eq. (3.16):

$$\overline{G}^{i}_{jk} = G^{i}_{jk} - (\dot{\partial}_{j}\dot{\partial}_{k}B^{im})\,\sigma_{m}. \quad (3.18)$$

Repeated transvections of this equation with directional coordinates also measures the variation in the conformal change of the functions G^{i}_{j} and G^{i}:

$$\overline{G}^{i}_{j} = G^{i}_{j} - (\dot{\partial}_{j}\,B^{im})\,\sigma_{m} \text{ (3.19a); and } \overline{G}^{i} = G^{i} - B^{im}\,\sigma_{m}, \text{ (3.19b)}$$

where relations (4.6.2a), (4.6.3) and homogeneous properties of the functions B^{im} are also used.

3.2. Conformal change in curvature tensor

Putting from Eqs. (3.18), (3.19) and using the homogenous properties of the tensor B^{im} the conformal change in the Berwald's tensors H^{i}_{j}, H^{i}_{jk} and H^{i}_{jkh} have been computed [102], Eqs. (2.4), (2.7) and (2.8):

$$\overline{H}^i_h = H^i_h + \sigma_m \{\dot{x}^r P_r (\dot{\partial}_h B^{im}) - 2P_h B^{im}\} + (\dot{\partial}_h B^{im})\dot{x}^r \; P_r \, \sigma_m$$

$$-2B^{im} P_h \, \sigma_m + \sigma_m \, \sigma_r \{2B^{sm} (\dot{\partial}_h \, \dot{\partial}_s B^{ir}) - (\dot{\partial}_h B^{sm}) \, \dot{\partial}_s B^{ir}, \qquad (3.20)$$

$$\overline{H}^i_{kh} = H^i_{kh} + 2\sigma_m \;\; P_{[k} \, (\dot{\partial}_{h]} \, B^{im}) - 2(\dot{\partial}_{[k} \, B^{im}) \, P_{h]} \, \sigma_m$$

$$+ 2\sigma_m \, \sigma_r (\dot{\partial}_{[k} B^{sm}) (\dot{\partial}_{h]} \, \dot{\partial}_s B^{ir}), \qquad (3.21)$$

$$\overline{H}^i_{khj} = H^i_{khj} + 2\sigma_m \, \dot{\partial}_j \, P_{[k} \, (\dot{\partial}_{h]} \, B^{im}) - 2\,(\dot{\partial}_j \, \dot{\partial}_{[k} \, B^{im}) \, P_{h]} \, \sigma_m$$

$$+ 2\sigma_m \, \sigma_r \, \dot{\partial}_j \, \{(\dot{\partial}_{[k} B^{sm}) (\dot{\partial}_{h]} \, \dot{\partial}_s B^{ir})\} + 2\sigma_r (\dot{\partial}_{[k} B^{im}) G^r_{h]mj}. \qquad (3.22)$$

Note 3.2. The variation in Berwald's curvature tensor obtained by Knebelman [59] and transcribed by Rund [189], Eq. (6.2.14) lacks the last two terms involving the tensors G^r_{hmj} in above equation. Prof. Rund gave his consent to the correction suggested by the author in a personal communication in 1966 (also cf. [102], p. 188).

In view of contractions of Berwald's curvature tensor and its associated tensors, defined by Eqs. (4.9.20), the variation in the contracted tensors is obtained by the author [102]. Further, the conformal variation in the projective curvature tensor and its associated tensors is also derived by therein; which are quite combersome.

3.3. Conformal connection parameters

Let us introduce a relative scalar Φ of weight $-2/n$ in terms of the metric function F of Finsler space F_n:

$$\Phi = F^2 \, | \, g \, |^{-1/n}, \qquad (3.23)$$

meeting all the conditions imposed on F defining a Finsler metric function (cf. Chapter 3, § 7) and a non-zero determinant $g \equiv | \, g_{ij} \, |$. Now, we construct a new set of functions G_{ij} in analogy with the metric tensor g_{ij} defined by Eq. (3.7.8):

$$G_{ij}(x^k, \, \dot{x}^k) \equiv (1/2) \, \dot{\partial}_i \, \dot{\partial}_j \, \Phi. \qquad (3.24)$$

These functions constitute a relative tensor of weight $-2/n$ and satisfy the coordinate transformation law

$$\overline{G}_{ab} = \left|\frac{\partial x^i}{\partial \overline{x}^a}\right|^{-2/n} . G_{ij} \left(\frac{\partial x^i}{\partial \overline{x}^a}\right) . \left(\frac{\partial x^j}{\partial \overline{x}^b}\right), \qquad (3.25)$$

The normalized cofactors G^{ij} of G_{ij} in det. $|G_{ij}|$ satisfy the coordinate transformation law

$$\overline{G}^{ab} = \left|\frac{\partial x^i}{\partial \overline{x}^a}\right|^{2/n} . G^{ij} \left(\frac{\partial \overline{x}^a}{\partial x^i}\right) . \left(\frac{\partial \overline{x}^b}{\partial x^j}\right), \qquad (3.26)$$

Now, beginning with the (new) metric tensor G_{ij} in the conformal Finsler space, we construct the generalized Christoffel symbols, in analogy with those defined by Eq. (4.3.1) and (4.3.2), for these functions

(a) $\lambda_{ikj} \equiv \partial_{(i} G_{j)k} - (1/2)\, \partial_k G_{ij}$, (b) $\lambda^i_{jk} \equiv G^{ih} \lambda_{jhk}$. (3.27)

The first kind symbols are symmetric in their first and last indices while the second kind symbols are symmetric in their lower indices. In analogy with Eq. (3.15), we now define the conformal connection parameters

$$\Lambda^i_{jk} = \lambda^i_{jk} + 2\,\dot{x}^l\, \dot{\partial}_{(j}\, \lambda^i_{k)l} + (1/2)\dot{x}^l\, \dot{x}^m\, \dot{\partial}_j\, \dot{\partial}_k\, \lambda^i_{lm}. \qquad (3.28)$$

Naturally, the coordinate transformation law for these new connection parameters is not the same as for G^i_{jk}.

Note 3.3. Conformal variation in the curvature tensors K^i_{jkh} and S^i_{jkh}, defined in Chapter 4, § 7, are also studied and the generalizations of Bianchi identities satisfied by the transformed curvature tensors $\overline{K}^i_{jkh}$, $\overline{S}^i_{jkh}$ and $\overline{H}^i_{jkh}$ have been derived by the author [102].

Note 3.4. The projective connection parameters defined by Eq. (1.13) have been seen invariant under the projective change. Their variation under the conformal change $\mathcal{C}$ is evaluated by the author [106], Eq. (2.3):

$$\overline{\Pi}^i_{jk} = \Pi^i_{jk} - \sigma_m(\dot{\partial}_j\, \dot{\partial}_k\, B^{im})$$

$$+\sigma_m\left\{2\,\delta^i_{(j}\, \dot{\partial}_{k)}\, \dot{\partial}_r B^{rm} + \dot{x}^i\, \dot{\partial}_r \dot{\partial}_j\, \dot{\partial}_k\, B^{rm}\right\}/\,(n+1)]. \qquad (3.29)$$

Such invariants have been called *projective invariants* in a conformal space $\overline{F}_n$. The projective covariant derivation of a geometric object w.r.t. projective connection parameters, introduced vide Eq. (1.20), also remains invariant under the projective change, if the object itself is a projective invariant. Notably, the geometric objects F, l^i, g_{ij}, C_{ijk}, A_{ijk}, C^i_{jk}, etc. are such (projective) invariants. So, their projective covariant derivatives also remain invariant under the projective change. Their conformal variation is also derived by the author [106].

§ 4. Infinitesimal transformation

Let the space F_n experience a point change of coordinates $x^i \to \bar{x}^i$, called an infinitesimal transformation:

$$\eta : \bar{x}^i = x^i + \varepsilon\, v^i(x^j), \tag{4.1}$$

where ε is an infinitesimal constant and $v^i(x^j)$ form the components of contravariant vector independent of directional arguments. The vector is called the *generator of the transformation*. Differentiating Eq. (4.1) partially with respect to x^j, we find

$$\partial \bar{x}^i / \partial x^j = \partial x^i / \partial x^j + \varepsilon\, (\partial v^i / \partial x^j) = \delta^i_j + \varepsilon\, (\partial_j v^i). \tag{4.2}$$

Further derivation of Eq. (4.2) gives the second order derivatives of $\bar{x}^i$:

$$\partial^2 \bar{x}^i / \partial x^j\, \partial x^k = \varepsilon\, (\partial^2 v^i / \partial x^j\, \partial x^k). \tag{4.3}$$

To get the inverse of above transformation, Eq. (4.1) is re-written as:

$$x^i = \bar{x}^i - \varepsilon\, v^i(x^j). \tag{4.4}$$

Its derivatives w.r.t. $\bar{x}^j$ are given by

$$\partial x^i / \partial \bar{x}^j = \delta^i_j - \varepsilon\, (\partial v^i / \partial x^k)\, (\partial x^k / \partial \bar{x}^j) \tag{4.5}$$

$$= \delta^i_j - \varepsilon\, (\partial_k v^i) \delta^k_j = \delta^i_j - \varepsilon\, (\partial_j v^i), \tag{4.6}$$

where substitutions for $\partial x^k / \partial \bar{x}^j$ are made from Eq. (4.5) and approximations are taken up to the first order of the infinitesimal constant ε. Accordingly, the second derivative of Eq. (4.6) reads as follows:

$$\partial^2 x^i / \partial \bar{x}^j\, \partial \bar{x}^k = -\, \varepsilon (\partial^2 v^i / \partial x^j\, \partial x^h)\, (\partial x^h / \partial \bar{x}^k) = -\, \varepsilon\, (\partial^2 v^i / \partial x^j\, \partial x^k), \tag{4.7}$$

again by Eq. (4.5) and retaining the terms in ε up to the first order.

Derivation of Eq. (4.1) w.r.t. the parameter t also determines the corresponding variation in the directional arguments:

$$\dot{\bar{x}}^i = \dot{x}^i + \varepsilon(\partial_j v^i)\dot{x}^j. \tag{4.8}$$

§ 5. Infinitesimal motion. Killing equations

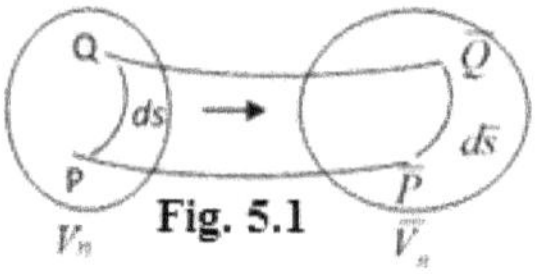

Fig. 5.1

Definition 5.1. Distance preserving infinitesimal transformations are called *motions*.

Let the infinitesimal transformations in Eq. (4.1) map points P (x^i) and Q $(x^i + dx^i)$ of the space F_n onto $\bar{\text{P}}$ $(\bar{x}^i)$ and $\bar{\text{Q}}$ $(\bar{x}^i + d\bar{x}^i)$ of $\bar{F}_n$ respectively. The arcual distances PQ and $\overline{\text{PQ}}$, denoted by ds and $d\bar{s}$, are respectively given by

$$(ds)^2 = g_{ij}(x^k, \dot{x}^k)dx^i dx^j, \quad \text{and} \quad (d\bar{s})^2 = \bar{g}_{ij}(\bar{x}^k, \dot{\bar{x}}^k)\, d\bar{x}^i d\bar{x}^j. \tag{5.1}$$

Putting from Eqs. (4.1) and (4.8), the last relation assumes the form

$$(d\bar{s})^2 = \bar{g}_{ij}\{x^k + \varepsilon v^k, \dot{x}^k + \varepsilon(\partial_h v^k)\dot{x}^h\}\{d(x^i + \varepsilon v^i)\}\{d(x^j + \varepsilon v^j)\}$$

$$= \{g_{ij}(x^k, \dot{x}^k) + \varepsilon(\partial_k g_{ij})v^k + 2\varepsilon C_{ijk}(\partial_h v^k)\dot{x}^h\}(dx^i + \varepsilon\, dv^i)(dx^j + \varepsilon\, dv^j)$$

$$= g_{ij}(x^k, \dot{x}^k)\, dx^i dx^j + \varepsilon\{g_{ij}\, dv^i dx^j + g_{ij}\, dx^i dv^j + (\partial_k g_{ij})v^k dx^i dx^j$$

$$+ 2C_{ijk}(\partial_h v^k)\dot{x}^h\, dx^i dx^j\}$$

$$= (ds)^2 + \varepsilon\{g_{kj}(\partial_i v^k) + g_{ik}(\partial_j v^k) + (\partial_k g_{ij})v^k$$

$$+ 2C_{ijk}(\partial_h v^k)\dot{x}^h\}dx^i dx^j, \tag{5.2}$$

by Taylor's theorem applied to the functions $\bar{g}_{ij}$, Eq. (5.1) and retaining the terms in ε up to the first order only. Thus, the necessary and sufficient conditions for an infinitesimal motion result from Eq. (5.2):

$$\{g_{kj}(\partial_i v^k) + g_{ik}(\partial_j v^k) + (\partial_k g_{ij})v^k + 2C_{ijk}(\partial_h v^k)\dot{x}^h\}\, dx^i dx^j = 0, \tag{5.3}$$

for arbitrary ε. The sum of the terms within crooked brackets in above equation is symmetric in the indices i and j, there results the condition

$$h_{ij} \equiv g_{kj}\,(\partial_i\, v^k) + g_{ik}\,(\partial_j\, v^k) + (\partial_k\, g_{ij})\, v^k + 2C_{ijk}\,(\partial_h v^k)\,\dot{x}^h = 0. \quad (5.4)$$

Note 5.1. Above equations representing the necessary and sufficient conditions for an infinitesimal motion are the generalized *Killing equations*.

Applying the covariant derivation process, given by Eq. (4.5.14), and noting the independence of the vector v^k on directional arguments, we derive

$$\partial_i\, v^k = \nabla_i\, v^k - v^h\, \Gamma^{*k}_{ih} \;. \quad (5.5a)$$

Transvecting it by g_{kl}, noting Eqs. (4.5.15) and introducing the associate of vector v^k:

$$g_{kl}\, v^k = v_l \quad (5.6); \quad \text{and} \quad g_{kl}\, \Gamma^{*k}_{ih} = \Gamma^{*}_{ilh}, \quad (5.7)$$

Eq. (5.5a) takes the form

$$g_{kl}\,(\partial_i\, v^k) = \nabla_i\, v_l - v^h \Gamma^{*}_{ilh}. \quad (5.5b)$$

Similarly, other terms in Eq. (5.4) get amended, and after simplification the equation transforms to

$$\nabla_i\, v_j + \nabla_j\, v_i + 2C_{ijk}(\nabla_h\, v^k)\,\dot{x}^h = 0. \quad (5.8)$$

5.1. Applications of Killing equation

From Eq. (4.1) we notice that the coordinates x^i's undergo the infinitesimal change

$$\delta x^i \equiv \bar{x}^i - x^i = \varepsilon\, v^i, \quad (5.9)$$

while the directional coordinates $\dot{x}^i$'s experience the change

$$\dot{\bar{x}}^i = \dot{x}^i + \varepsilon\,(\partial_j\, v^i)\,\dot{x}^j. \quad (5.10)$$

So, the magnitude of the infinitesimal displacement $\mathrm{P\bar{P}}$ in a motion is given by

$$(\delta s)^2 = g_{ij}\,\delta x^i\,\delta x^i = \varepsilon^2 g_{ij}\, v^i\, v^j \;. \quad (5.11)$$

Therefore, for a non-null motion we must have

$$g_{ij}\, v^i\, v^j \neq 0. \quad (5.12)$$

Thus, when the Killing Eqs. (5.4) are consistent and admit a solution satisfying Eq. (5.12) the vectors v^i' s determine the infinitesimal generator of a group G_r of motions of F_n. Choosing the coordinates x^i's in a way so that the vectors v^i's have components δ^i_1, i.e.

$$v^1 = 1, \qquad v^j = 0 \Rightarrow \qquad \partial_k v^i = 0, \qquad (5.13)$$

for every i, k and $j = 2, 3, \ldots, n$. Accordingly, the infinitesimal transformations in Eq. (4.1) reduce to

$$\bar{x}^1 = x^1 + \varepsilon, \quad \bar{x}^j = x^j \quad (j = 2, 3, \ldots, n); \qquad (5.14)$$

and the Killing equations (5.4) simplify as

$$\partial g_{ij} / \partial x^1 = 0, \quad (i, j = 1, 2, 3, \ldots, n). \qquad (5.15)$$

This results in the independence of the functions g_{ij} on the coordinate x^1. The infinitesimal displacement δx^i given by Eq. (5.10), for Eqs. (5.14), reduces to

$$\delta x^1 = \varepsilon, \quad \delta x^j = 0 \ \ (j = 2, 3, \ldots, n). \qquad (5.16)$$

Consequently, the fundamental form in Eq. (5.11) is transformed as

$$(\delta s)^2 = \varepsilon^2.\{g_{11}\, v^1 v^1 + g_{1j}\, v^1 v^j + g_{j1}\, v^j v^1 + g_{ij}\, v^i v^j\}, \qquad (5.17)$$

where both i and j run from 2 to n. For Eq. (5.13), above relation further simplifies to

$$(\delta s)^2 = g_{11}\, \varepsilon^2, \qquad (5.18)$$

As per Eq. (5.15), g_{11} is independent of the coordinate x^1 and ε is an infinitesimal constant so Eq. (5.18) shows that the fundamental form is transformed onto itself by Eq. (5.13). So, we have the:

Theorem 5.1. If a space F_n admits an infinitesimal motion it admits the finite continuous group G_1 of motions generated by the infinitesimal motion.

Conversely, when Eqs. (5.15) are satisfied, a solution of Killing equations (5.4) is given by Eq. (5.15). Therefore, we also have the:

Theorem 5.2. A necessary and sufficient condition that a F_n admits an infinitesimal motion is that there exists a coordinate system for which g_{ij}'s are independent of one of the coordinates, say x^1. The parametric curves x^1 are the paths of infinitesimal motion.

Theorem 5.3. A motion preserves the length of the infinitesimal vector dx^i.

Proof. The operator δ of the infinitesimal change in Eq. (5.9) measures the change in dx^i by

$$\delta\,(dx^i) = d\,(\delta x^i) = d\,(\varepsilon v^i) = \varepsilon\,(dv^i) = \varepsilon\,(\partial_k\, v^i)\, dx^k. \qquad (5.19)$$

On the other hand, the infinitesimal change in the metric tensor g_{ij} is measured by

$$\delta g_{ij} = (\partial_k\, g_{ij})\,\delta x^k + (\dot{\partial}_k\, g_{ij})\,\delta \dot{x}^k = \varepsilon\{\, v^k\,\partial_k\, g_{ij} + 2C_{ijk}(\partial_h\, v^k)\,\dot{x}^h\,\}, \quad (5.20)$$

by Eqs. (5.9) and (5.10). Therefore, the infinitesimal change in the magnitude of dx^i is given by

$$\delta\,(ds)^2 = \delta\,(g_{ij}\,dx^i\,dx^j) = (\delta\, g_{ij})\,dx^i\,dx^j + g_{ij}\,\{\delta\,(dx^i)\}\,dx^j$$

$$+\, g_{ij}\,dx^i\,\{\delta\,(dx^j)\} = \varepsilon\,\{\, v^k\,\partial_k\, g_{ij} + 2C_{ijk}\,\dot{x}^h\,\partial_h\, v^k\,\}\,dx^i\,dx^j$$

$$+\, g_{ij}\,(\partial_k\, v^i)\,dx^k\,dx^j + g_{ij}\;dx^i\;(\partial_k\, v^j)\,dx^k\,\}$$

$$= \varepsilon\,\{\, v^k\,\partial_k\, g_{ij} + g_{aj}\,(\partial_i\, v^a) + g_{ia}\,(\partial_j\, v^a) + 2C_{ijk}\,\dot{x}^h\,\partial_h\, v^k\,\}\,dx^i\,dx^j = 0,$$

by Eq. (5.4) holding for infinitesimal motion. This establishes the invariance of $(ds)^2$ under infinitesimal motion defined by Eq. (4.1). //

Theorem 5.4. An infinitesimal motion preserves the angle between two directions.

Proof. Let $d_1\, x^i$ and $d_2\, x^i$ be any two directions in F_n inclined at an angle θ at some point P. So that we have

$$\cos\theta = g_{ij}\,d_1\, x^i\, d_2\, x^j \,/\, \surd\,(g_{ab}\, d_1 x^a\, d_1\, x^b\,).\surd\,(g_{ce}\, d_2\, x^c\, d_2\, x^e). \qquad (5.21)$$

The effect of infinitesimal change on $g_{ij}\, d_1 x^i\, d_2 x^j$ is measured by

$$\delta\,\{g_{ij}\,(d_1\, x^i)(d_2\, x^j)\} = \varepsilon\,[\{v^k\,(\partial_k\, g_{ij}) + 2C_{ijk}(\partial_h\, v^k)\,\dot{x}^h\,\}(d_1\, x^i)(d_2\, x^j)$$

$$+\, g_{ij}\,(\partial_k\, v^i\,)\,(d_1\, x^k)\,(d_2\, x^j) + g_{ij}\,(d_1\, x^i)\,(\partial_k\, v^j)\,(d_2 x^k)\,]$$

$$= \varepsilon\,\{v^k\,\partial_k\, g_{ij} + g_{aj}\,(\partial_i\, v^a) + g_{ia}\,(\partial_j\, v^a) + 2C_{ijk}\,(\partial_h\, v^k)\,\dot{x}^h\,\}d_1 x^i\, d_2 x^j, \quad (5.22)$$

by Eqs. (5.19) and (5.20). The same, in view of Eq. (5.4), vanishes for a motion. The factors with radical signs in Eq. (5.21) measure the magnitudes of infinitesimal vectors $d_1 x^i$ and $d_2 x^i$ which remain invariant under an infinitesimal motion as per Theorem 5.3. Hence, operating Eq. (5.21) by δ, we obtain δ (cos θ) = 0, establishing the theorem. //

§ 6. Infinitesimal translation

Definition 6.1. An infinitesimal motion carrying each point P of F_n through the same distance is called an *infinitesimal translation*.

Thus, we observe from Eq. (5.11) that

$$g_{ij} v^i v^j = \text{constant} \tag{6.1}$$

for an infinitesimal translation. Indeed, Eq. (6.1) represents a necessary and sufficient condition for an infinitesimal motion to define an infinitesimal translation.

If the generators of an infinitesimal motion are taken as per Eq. (5.13) giving rise to Eqs. (5.15), the condition vide Eq. (6.1) reduces to g_{11} = const.

§ 7. Infinitesimal conformal motion. Homothetic motion

The conformal transformations are characterized by Eq. (3.2b). Let $\overline{\theta}$ be the angle between the images of two directions $d_1 \overline{x}^i$ and $d_2 \overline{x}^i$ in $\overline{F}_n$. Analogous to Eq. (5.21), this angle is given by

$$\cos \overline{\theta} = \overline{g}_{ij} d_1 x^i d_2 x^j / \surd(\overline{g}_{ab} d_1 x^a d_2 x^b).\surd(\overline{g}_{ce} d_2 x^c d_2 x^e)$$

$$= e^{2\sigma} g_{ij}\, d_1 x^i d_2 x^j / e^{2\sigma} \surd(g_{ab}\, d_1 x^a d_1 x^b).\surd(g_{ce}\, d_2 x^c d_2 x^e), \quad \text{by Eq. (3.2b)}$$

$$= \cos\theta \Rightarrow \overline{\theta} = \theta,$$

by Eq. (5.21) again. Thus, the angle between corresponding directions in the two spaces remains constant under a conformal transformation.

Note 7.1. It may be noted that under a conformal transformation of F_n onto $\overline{F}_n$ only the metric tensors of the two spaces experience the change and not the directions dx^i's.

The infinitesimal change in $g_{ij}\ d_1 x^i\ d_2 x^j$ has been derived in Eq. (5.22):

$$\delta\,\{g_{ij}\,(d_1 x^i)(d_2 x^j)\} \equiv \varepsilon\, h_{ij}\ (d_1 x^i)(d_2 x^j), \qquad (7.1)$$

where h_{ij} are given by Eq. (5.4). In view of above Note, Eq. (7.1) may be written as

$$(\delta\, g_{ij} - \varepsilon\, h_{ij})\, d_1 x^i\ d_2 x^j\ = 0 \quad \Rightarrow \quad \delta\, g_{ij} = \varepsilon\, h_{ij}\,, \qquad (7.2)$$

for arbitrary directions. But, $\delta\, g_{ij} = \overline{g}_{ij} - g_{ij}$, Eq. (7.2) yields

$$\overline{g}_{ij} = g_{ij} + \varepsilon\, h_{ij}. \qquad (7.3)$$

Comparing Eqs. (3.2b) and (7.3) we obtain

$$\varepsilon\, h_{ij}\ =\ (e^{2\sigma} - 1)\, g_{ij}\,, \qquad \text{or,} \qquad h_{ij}\ =\ \psi\, g_{ij}\,, \qquad (7.4)$$

for suitable choice of the scalar function ψ. The Eq. (7.4) characterizes an infinitesimal conformal transformation. When the function ψ remains constant the conformal transformation is called *homothetic*. Moreover, when $\psi = 0$ causing $h_{ij} = 0$, the transformations defines an infinitesimal motion as characterized by Eq. (5.4).

Thus, an infinitesimal motion and homothetic transformations are special cases of conformal transformations.

§ 8. Infinitesimal transformations preserving geodesics

It is seen in § 1 that the infinitesimal transformations given by Eq. (4.1) map a geodesic of F_n onto a geodesic of $\overline{F}_n$ if the connection parameters of two spaces are connected by Eq. (1.7) for a suitable choice of the scalar functions p satisfying Eqs. (1.9). We shall deal with this topic in detail, later, in the next chapter.

§ 9. Infinitesimal transformations generated by vector field of diverse choices

For Riemannian case Takano [220-V] discusses various possibilities for the vector-field $v^i\ (x^j)$ generating the infinitesimal transformations given by Eq. (4.1). Various authors including the present one studied such infinitesimal transformations in different Finslerian structures [78], [85], [96], [108], [111] - [113], [120], [121], [132], [134], [135], [137], [166], [174], [175], [179], [181], etc.

Definition 9.1. The following conditions imposed on the choice of vector field $v^i\,(x^j)$ rise to the different kind of transformations indicated therein:

(a) $\mathfrak{B}_j\, v^i = 0$: Contra vector field; (9.1a)

(b) $\mathfrak{B}_j\, v^i = c\ \delta^i_j$, $c \neq 0$ constant (a real number): (9.1b)

(Concurrent vector field);

(c) $\mathfrak{B}_j\, v^i = \rho\ (x^k, \dot{x}^k)\ \delta^i_j,\ \ \rho \neq 0$: (9.1c)

(Special concircular vector field);

(d) $\mathfrak{B}_j\, v^i = \mu_j\ (x^k, \dot{x}^k)\ v^i,\ \ \mu_j \neq 0$: Recurrent vector field; (9.1d)

(e) $\mathfrak{B}_j\, v^i = \rho\ (x^k, \dot{x}^k)\ \delta^i_j + \mu_j\ (x^k, \dot{x}^k) v^i,\qquad \mu_j \neq 0,$ (9.1e)

satisfying $2\mathfrak{B}_{[j}\ \mu_{k]} = 0$: Concircular vector field; (9.2)

(f) $\mathfrak{B}_j\, v^i = \rho\ (x^k, \dot{x}^k)\ \delta^i_j + \mu_j\ (x^k, \dot{x}^k) v^i,\ \mu_j \neq 0$: (9.1f)

(Torse-forming vector field).

Note 9.1. $\mathfrak{B}_j\, v^i$ being positively homogeneous of degree zero in the directional arguments, both the scalar function ρ and the recurrent vector field μ_j are also of the same nature.

Hence, the second order covariant derivatives of the vector field v^i are:

$$\mathfrak{B}_j\, \mathfrak{B}_k\, v^i = 0 \text{ (for both contra and concurrent)}; \tag{9.3}$$

$$\mathfrak{B}_j\, \mathfrak{B}_k\, v^i = \rho_j \delta^i_k\,,\qquad \rho_j \equiv \mathfrak{B}_j\, \rho \text{ (for special concircular)}; \tag{9.4}$$

$$\mathfrak{B}_j\, \mathfrak{B}_k\, v^i = (\mathfrak{B}_j\, \mu_k + \mu_j\, \mu_k)\, v^i = \mu_{jk} v^i \text{ (for recurrent vector)}, \tag{9.5}$$

where

$$\mu_{jk} \equiv \mathfrak{B}_j\, \mu_k + \mu_j\, \mu_k; \tag{9.6}$$

$$\mathfrak{B}_j\, \mathfrak{B}_k\, v^i = \rho_j\, \delta^i_k + \rho\, \delta^i_j\, \mu_k + \mu_{jk} v^i \text{ (for concircular vector)}, \tag{9.7}$$

where μ_{jk} is symmetric tensor; and

$$\mathfrak{B}_j\, \mathfrak{B}_k\, v^i = \rho_j\, \delta^i_k + \rho\, \delta^i_j\, \mu_k + \mu_{jk} v^i \text{ (for torse-forming vector)}. \tag{9.8}$$

9.1. F_n with special concircular infinitesimal transformations

The commutation formula in Eq. (4.8.3), when applied to the vector v^i (independent of $\dot{x}^i$'s), for Eq. (9.1c), reduces to

$$(\dot{\partial}_k \rho)\, \delta^i_h = v^j\, G^i_{jkh}. \tag{9.9}$$

On transvection with $\dot{x}^h$, for Eq. (4.10.15), it yields

$$\dot{x}^i\, (\dot{\partial}_k \rho) = 0 \quad \Rightarrow \quad \dot{\partial}_k \rho = 0, \tag{9.10}$$

making ρ a point function. Hence, it follows from Eq. (4.6.6) that the vector given by Eq. (9.4):

$$\rho_k \equiv {}_k \rho = \partial_k \rho, \tag{9.11a}$$

is a gradient vector independent of $\dot{x}^i$'s and it implies

$$2\mathfrak{B}_{[j}\, \rho_{k]} = 0, \quad \text{i.e.} \quad \mathfrak{B}_j\, \rho_k = \mathfrak{B}_k\, \rho_j. \tag{9.11b}$$

Also, from Eqs. (9.9) and (9.10), there follows the relation

$$v^j\, G^i_{jkh} = 0. \tag{9.12}$$

On the other hand, the commutation formula in Eq. (4.7.14), when applied to the vector v^i satisfying Eq. (9.1c), yields

$$2\rho_{[j}\, \delta^i_{k]} = H^i_{jkh}\, v^h. \tag{9.13}$$

Contracting it for the indices i and j, and putting from Eq. (4.9.20), it determines the vector

$$\rho_k = H_{kh}\, v^h /(1-n). \tag{9.14}$$

Eliminating ρ_k from the last two equations there follows

$$H^i_{jkh}\, v^h = 2\,\delta^i_{[j}\, H_{k]h}\, v^h /(n-1), \tag{9.15a}$$

i.e.

$$v^h\, \{H^i_{jkh} - 2\,\delta^i_{[j}\, H_{k]h} /(n-1)\} = 0, \tag{9.15b}$$

Its covariant derivation, for Eq. (9.1c), yields

$$(v^h\, \mathfrak{B}_l + \rho\, \delta^h_l)\{H^i_{jkh} - 2\,\delta^i_{[j}\, H_{k]h} /(n-1)\} = 0. \tag{9.16}$$

We shall continue this discussion further in Chapter 9 dealing with a space of recurrent curvature.

Note 9.2. For Eq. (9.11a), the vector ρ_k is a point function and so are the tensor in Eq. (9.13) and the vector-field in Eq. (9.14).

9.2. Concircular infinitesimal transformations (c.i.t.) in F_n

Beginning with Eq. (9.1e) and proceeding similarly, the author et al. [135] obtained more general results. Thus, Eq. (9.13) and (9.14) read as

$$2\,\delta^i_{[j}\,\{\rho\,\mu_{k]} - \rho_{k]}\} = H^i_{jkh}\,v^h, \tag{9.17}$$

and

$$(n-1)(\rho\,\mu_k - \rho_k) = H_{kh}\,v^h. \tag{9.18}$$

Eliminating $\rho\,\mu_k - \rho_k$ from the last two equations there again follows Eq. (9.15a). Similarly, Eq. (9.9), for c.i.t. reads as

$$(\dot{\partial}_k\,\rho)\,\delta^i_h + (\dot{\partial}_k\,\mu_h)\,v^i = v^j\,G^i_{jkh}. \tag{9.19}$$

Its skew-symmetric part w.r.t. indices k and h gives

$$2\,(\dot{\partial}_{[k}\,\rho)\,\delta^i_{h]} + 2\,v^i(\dot{\partial}_{[k}\,\mu_{h]}) = 0, \tag{9.20}$$

which, on transvection with v^k, yields

$$v^k\,(\dot{\partial}_k\,\rho)\,\delta^i_h - v^i\,\{\dot{\partial}_h\,\rho - 2\,v^k\,\dot{\partial}_{[k}\,\mu_{h]}\} = 0. \tag{9.21}$$

Further, its contraction w.r.t. indices i and h determines

$$(n-1)\,v^i\,\dot{\partial}_i\,\rho + 2\,v^i\,v^k\,\dot{\partial}_{[k}\,\mu_{i]} = 0. \tag{9.22}$$

The second term being zero, it reduces to

$$v^i\,\dot{\partial}_i\,\rho = 0. \tag{9.23}$$

Accordingly, for arbitrary v^i, Eq. (9.21), yields

$$\dot{\partial}_h\,\rho - 2\,v^k\,\dot{\partial}_{[k}\,\mu_{h]} = 0. \tag{9.24}$$

On the other hand, contraction of Eq. (9.20) w.r.t. indices i and k yields

$$(1-n)\dot{\partial}_h \rho + 2 v^k \dot{\partial}_{[k} \mu_{h]} = 0. \tag{9.25}$$

Addition of the last two equations, thus, proves the independence of ρ on $\dot{x}^i$'s in an F_n $(n > 2)$ as in Sub-section 9.1, Eq. (9.10). As such, Eq. (9.20) reduces to

$$2(\dot{\partial}_{[k} \mu_{h]}) = \dot{\partial}_k \mu_h - \dot{\partial}_h \mu_k = 0 \quad \Rightarrow \quad \mu_k = \dot{\partial}_k \mu \tag{9.26}$$

for some scalar function μ. Thus, we have the

Theorem 9.1. In a Finsler space F_n $(n > 2)$ admitting a c.i.t. the function ρ is necessarily a point function while the function μ satisfies Eq. (9.26).

Note 9.1. As seen in Eq. (9.10), the function ρ is independent of the directional arguments and above theorem also holds good for special concircular infinitesimal transformations in F_n.

9.3. Special c.i.t. in F_n with $\rho_k = 0$

In view of Eq. (9.10) and above Note, ρ is a point function. Thus, the said hypothesis $\rho_k = 0$, for Eq. (9.11) turns ρ merely a constant, say c. Accordingly, the transformation reduces to *contra* (in case $c = 0$) else a *concurrent* one. The equations (9.13) and (9.14), thus, reduce to

$$H^i_{jkh} v^h = 0 \quad (9.27); \quad \text{and} \quad H_{kh} v^h = 0, \tag{9.28}$$

respectively. Further, the equations (9.15) trivially hold in case of contra or concurrent c.i.t.'s. It may be noted that there does not exist Eq. (9.16).

Study of concircular infinitesimal transformations for different kinds of Finslerian structures is continued in Chapters 12 and 13.

§ 10. Infinitesimal transformations generated by a recurrent vector field

Eq. (9.1d) represents the recurrent form of the vector field v^i generating an infinitesimal transformation in Eq. (4.1), where the vector-field μ_k is positively homogeneous function of degree zero in $\dot{x}^i$'s. The second order covariant derivative of v^i is found by Eq. (9.5). The skew-symmetric part of this derivative, for the commutation rule vide Eq. (4.7.14) applied to the vector v^i, yields

$$2\, v^i\, \mathfrak{B}_{[j}\, \mu_{k]} = H^i_{jkh}\, v^h. \tag{10.1}$$

On the other hand, the commutation formula vide Eq. (4.8.3), when applied to the vector v^i satisfying Eq. (9.1d), gives

$$v^i\, \dot{\partial}_j\, \mu_k = G^i_{jkh}\, v^h. \tag{10.2}$$

The tensor G^i_{jkh} being symmetric in its lower indices, the skew-symmetric part of above equation w.r.t. the indices j and k vanishes:

$$2v^i\, \dot{\partial}_{[j}\, \mu_{k]} = 0, \tag{10.3}$$

implying Eq. (9.26). Transvection of Eq. (9.26) with $\dot{x}^k$, for homogeneous properties of μ, determines the function μ:

$$\dot{x}^k\, \mu_k = \dot{x}^k\, \dot{\partial}_k\, \mu = \mu, \tag{10.4}$$

which, on differentiation w.r.t. $\dot{x}^h$, for Eq. (9.26), also yields

$$\dot{x}^k\, \dot{\partial}_h\, \mu_k = 0. \tag{10.5}$$

Hence, the function μ_{jk} defined by Eq. (9.6), for Eqs. (4.6.7) and (10.4), also satisfies

$$\mu_{jk}\, \dot{x}^k \equiv \mathfrak{B}_j\, \mu + \mu\, \mu_j. \tag{10.6}$$

Foot-note:

29) For Eulerian curvature tensors see Rund [189], pp. 165-169.

CHAPTER 6

THEORY OF LIE DERIVATIVES

§ 1. Introduction

W. Slebodzinski [216] introduced a new derivation process in 1931 that was later called *Lie derivation* by D. van Dantzig. Dantzig also applied the theory of Lie derivation to physics. Currently, we derive formulae for Lie derivatives of various types of geometric quantities such as scalar functions, vectors, tensors, connection parameters and projective connection parameters.

§ 2. Lie derivative

2.1. Lie derivative of a geometric object

Let $\Omega\ (x^k, \dot{x}^k)$ be a geometric object of the Finsler space F_n that gets transformed to $\overline{F}_n$ under an infinitesimal transformation defined by Eq. (5.4.1). Let Ω maps onto $\overline{\Omega}(\overline{x}^k, \dot{\overline{x}}^k)$ under the transformation and experiences the change (analogous the one seen in the functions $\overline{g}_{ij}$ in Eq. (5.5.2)):

$$d^v\,\Omega \equiv \overline{\Omega}(\overline{x}^k, \dot{\overline{x}}^k) - \Omega(x^k, \dot{x}^k) = (\partial_j\,\Omega)\,dx^j + (\dot{\partial}_j\,\Omega)\,d\dot{x}^k$$

$$= \varepsilon\{(\partial_j\,\Omega)\,v^j + (\dot{\partial}_j\,\Omega)(\partial_k v^j)\,\dot{x}^k\}\,, \tag{2.1}$$

by Eqs. (5.4.1), (5.4.8) and approximations in retaining the powers of ε up to first order. On the other hand, observing the infinitesimal transformation merely as a coordinate change, let the object Ω transforms to $\widetilde{\Omega}(\overline{x}^k, \dot{\overline{x}}^k)$ - called the '*displaced*' object [189] and experience the change

$$d^m\,\Omega \equiv \widetilde{\Omega}(\overline{x}^k, \dot{\overline{x}}^k) - \Omega(x^k, \dot{x}^k)\,. \tag{2.2}$$

Definition 2.1. The Lie derivative of the object Ω with respect to the infinitesimal transformation is defined by the limit

$$\pounds\,\Omega \equiv \lim_{\varepsilon\to 0}(d^v\,\Omega - d^m\,\Omega)/\varepsilon\,. \tag{2.3a}$$

Note 2.1. The notation used by Yano [234] for the operation of Lie derivation is retained here, while Rund [189] employed the notation $\underset{L}{D}$ for it.

Note 2.2. If Ω is a scalar function, i.e. it remains unaltered under the coordinate change causing $d^m\,\Omega = 0$ and it reduces the formula (2.3a):

$$\pounds\,\Omega \equiv \lim_{\varepsilon\to 0}(d^{\,v}\Omega)/\varepsilon = (\partial_j\,\Omega)\,v^j + (\dot\partial_j\,\Omega)(\partial_k v^j)\,\dot x^k, \quad (2.3b)$$

by Eq. (2.1). Changing the partial derivatives Ω and v^j into covariant derivatives by Eq. (4.5.14), above formula can be written as

$$\pounds\,\Omega = (\nabla_j\,\Omega)\,v^j + (\dot\partial_j\,\Omega)(\nabla_k v^j)\,\dot x^k, \quad (2.3c)$$

or, equivalently

$$\pounds\,\Omega = v^j\,(\mathfrak{B}_j\,\Omega) + (\dot\partial_j\,\Omega)\,(\mathfrak{B}_k\,v^j)\;\dot x^k. \quad (2.3d)$$

Example 2.1. Lie derivative of the metric function $F\,(x^k, \dot x^k)$ is

$$\pounds\,F = (\partial_j\,F)\,v^j + (\dot\partial_j\,F)(\partial_k\,v^j)\,\dot x^k = l_j\,(\nabla_k\,v^j)\,\dot x^k. \quad (2.4)$$

Solution. Writing the covariant derivative formula (4.5.14) for the scalar function F:

$$\nabla_j\,F = \partial_j\,F - (\dot\partial_k\,F)\,G^k_j, \quad (2.5)$$

and noting Eqs. (4.2.4), (4.5.16), (4.6.2b) and (5.5.5a), we derive the desired result from Eq. (2.3b). //

2.2. Lie derivative of a contravariant vector X^i

The vector transforms under the coordinate change given by Eq. (5.4.1) as

$$\widetilde X^i\,(\bar x^k, \dot{\bar x}^k) = X^j(x^k, \dot x^k)\,(\partial_j \bar x^i)$$
$$= X^j(\delta^i_j + \varepsilon\,\partial_j v^i) = X^i + \varepsilon\,X^j\,\partial_j\,v^i, \quad (2.6)$$

by Eq. (5.4.2). Hence, the change $d^m X^i$ is given by Eq. (2.2):

$$d^m X^i \equiv \widetilde X^i\,(\bar x^k, \dot{\bar x}^k) - X^i(x^k, \dot x^k) = \varepsilon\,X^j\,\partial_j\,v^i, \quad (2.7)$$

whereas the change $d^{\,v}X^i$ can be obtained from Eq. (2.1) applied for the vector X^i. Hence, the formula (2.3b) reads as

$$£X^i = (\partial_j X^i)v^j + (\dot{\partial}_j X^i)(\partial_k v^j)\dot{x}^k - X^j \partial_j v^i. \qquad (2.8a)$$

Putting from Eqs. (4.5.14), (5.5.5a), it can be transformed into a tensorial form

$$£X^i = v^j \nabla_j X^i - X^j \nabla_j v^i + (\dot{\partial}_j X^i)(\nabla_k v^j)\dot{x}^k. \qquad (2.8b)$$

Particularly, for Eqs. (4.2.7), (4.5.16), and independence of the vector v^k on directional arguments, there also follow the results

$$£v^i = 0 \qquad (2.9a); \qquad £\dot{x}^i = 0 \qquad (2.9b);$$

$$£l^i = (\delta^i_j - l^i l_j)(\nabla_k v^j) l^k. \qquad (2.9c)$$

Note 2.3. Employing Berwald's covariant derivation process given by Eq. (4.6.6), an alternate tensorial form of Eq. (2.8a) can be also similarly obtained:

$$£X^i = v^j(\mathfrak{B}_j X^i) - X^j(\mathfrak{B}_j v^i) + (\dot{\partial}_j X^i)(\mathfrak{B}_k v^j)\dot{x}^k. \qquad (2.8c)$$

2.3. Lie derivative of a covariant vector Y_i

The vector transforms under the coordinate change given by Eq. (5.4.1) as

$$\tilde{Y}_j(\bar{x}^k, \dot{\bar{x}}^k) = Y_i(x^k, \dot{x}^k)(\partial x^i / \partial \bar{x}^j)$$

$$= Y_i(\delta^i_j - \varepsilon \partial_j v^i) = Y_j - \varepsilon Y_i(\partial_j v^i), \qquad (2.10)$$

by Eq. (5.4.6). Hence, the change $d^m Y_j$, by Eq. (2.2), is:

$$d^m Y_j \equiv \tilde{Y}_j(\bar{x}^k, \dot{\bar{x}}^k) - Y_j(x^k, \dot{x}^k) = -\varepsilon Y_i(\partial_j v^i), \qquad (2.11)$$

The change $d^v Y_j$ can be obtained from Eq. (2.1) applied for the vector Y_j and $£Y_j$ is obtained from the formula (2.3a):

$$£Y_i = (\partial_j Y_i)v^j + (\dot{\partial}_j Y_i)(\partial_k v^j)\dot{x}^k + Y_j \partial_i v^j. \qquad (2.12a)$$

Applying the commutation formula (4.5.14) to the vector Y_j and putting from Eq. (5.5.5a), after cancellation of common terms with opposite signs, it transforms to a tensorial form:

$$\pounds\, Y_i = v^j \nabla_j Y_i + Y_j \nabla_i v^j + (\dot{\partial}_j Y_i)(\nabla_k v^j)\, \dot{x}^k; \qquad (2.12b)$$

or, equivalently

$$\pounds\, Y_i = v^j (\mathfrak{B}_j Y_i) + Y_j (\mathfrak{B}_i v^j) + (\dot{\partial}_j Y_i)(\mathfrak{B}_k v^j)\, \dot{x}^k. \qquad (2.12c)$$

Particularly, for Eqs. (4.2.8) and (4.5.16), we have

$$\pounds l_i = l_j \nabla_i v^j + (g_{ij} - l_i l_j)(\nabla_k v^j)\, l^k. \qquad (2.13)$$

2.4. Lie derivative of a mixed tensor T^i_j

Since the tensor T^i_j satisfies the coordinate transformation law

$$\widetilde{T}^i_j(\bar{x}^k, \dot{\bar{x}}^k) = T^a_b(x^k, \dot{x}^k)(\partial_a \bar{x}^i)(\partial x^b / \partial \bar{x}^j)$$

$$T^a_b(\delta^i_a + \varepsilon \partial_a v^i)(\delta^b_j - \partial_j v^b) = T^i_j + \varepsilon \{ T^a_j \partial_a v^i - T^i_b \partial_j v^b \}$$

$\Rightarrow$

$$d^m T^i_j \equiv \widetilde{T}^i_j(\bar{x}^k, \dot{\bar{x}}^k) - T^i_j(x^k, \dot{x}^k) = \varepsilon (T^a_j \partial_a v^i - T^i_b \partial_j v^b). \qquad (2.14)$$

Therefore, from Eqs. (2.1), (2.3a) and (2.14), we deduce

$$\pounds T^i_j = (\partial_k T^i_j)\, v^k + (\dot{\partial}_k T^i_j)(\partial_h v^k)\, \dot{x}^h - T^a_j \partial_a v^i + T^i_a \partial_j v^a. \quad (2.15)$$

Putting from Eqs. (4.5.14), (5.5.5a), after cancellation of common terms with opposite signs, it transforms to a tensorial form:

$$\pounds T^i_j = v^k \nabla_k T^i_j - T^k_j \nabla_k v^i + T^i_k \nabla_j v^k + (\dot{\partial}_k T^i_j)(\nabla_h v^k)\, \dot{x}^h, \quad (2.16a)$$

or, equivalently

$$\pounds T^i_j = v^j \mathfrak{B}_j T^i_j - T^k_j \mathfrak{B}_k v^i + T^i_k \mathfrak{B}_j v^k + (\dot{\partial}_k T^i_j)(\mathfrak{B}_h v^k)\, \dot{x}^h. \quad (2.16b)$$

2.5. For a tensor of arbitrary rank, say (r, s), the formulae (2.16) can be generalized as follows:

$$\pounds T^{i_1 i_2 \ldots i_r}_{j_1 j_2 \ldots j_s} = v^k \nabla_k T^{\cdots}_{\cdots} - \sum_{p=1}^{r} T^{i_1 i_2 \ldots i_{p-1} k\, i_{p+1} \ldots i_r}_{j_1 j_2 \ldots j_s} \nabla_k v^{i_p}$$

$$+ \sum_{q=1}^{s} T^{i_1 i_2 \ldots i_r}_{j_1 j_2 \ldots j_{q-1} k\, j_{q+1} \ldots j_s} \nabla_{j_q} v^k + (\dot{\partial}_k T^{\cdots}_{\cdots})(\nabla_h v^k)\, \dot{x}^h. \quad (2.17)$$

Theorem 2.1. Operation of Lie derivation is linear:

$$\pounds(X^i Y_j) = (\pounds X^i) Y_j + X^i (\pounds Y_j), \tag{2.18}$$

and

$$\pounds(a X^i \pm b Y^i) = a(\pounds X^i) + b(\pounds Y^i), \tag{2.19}$$

for any vectors $\mathbf{X}$, $\mathbf{Y}$ and constant scalars *a*, *b*.

Proof. (i) $X^i Y_j$ being a mixed tensor, its Lie derivative may be found by Eq. (2.16a):

$$\pounds(X^i Y_j) = v^k \nabla_k (X^i Y_j) - (X^k Y_j) \nabla_k v^i + (X^i Y_k) \nabla_j v^k$$
$$+ \{\dot{\partial}_k (X^i Y_j)\} (\nabla_h v^k) \dot{x}^h. \tag{2.20}$$

For linear properties of the operators ∇_k and $\dot{\partial}_k$, RHS of above equation splits as

$$\{v^k \nabla_k X^i - X^k \nabla_k v^i + (\dot{\partial}_k X^i)(\nabla_h v^k) \dot{x}^h\} Y_j$$
$$+ X^i \{v^k \nabla_k Y_j + Y_k \nabla_j v^k + (\dot{\partial}_k Y_j)\} (\nabla_h v^k) \dot{x}^h,$$

which, by Eqs. (2.8b) and (2.12b), is same as the RHS of Eq. (2.18).

(ii) Again, the linear sum $a X^i \pm b Y^i$ of two (same order) vectors being a vector of the same order, its Lie derivative may be found by Eq. (2.8b):

$$\pounds(a X^i \pm b Y^i) = v^j \nabla_j (a X^i \pm b Y^i) - (a X^j \pm b Y^j) \nabla_j v^i$$
$$+ \{\dot{\partial}_j (a X^i \pm b Y^i)\} (\nabla_k v^j) \dot{x}^k. \tag{2.21}$$

For the linear properties of the operators ∇_k and $\dot{\partial}_k$, RHS of above equation splits as

$$a\{v^j \nabla_j X^i - X^j \nabla_j v^i + (\dot{\partial}_j X^i)(\nabla_k v^j) \dot{x}^k\}$$
$$\pm b\{v^j \nabla_j Y^i - Y^j \nabla_j v^i + (\dot{\partial}_j Y^i)(\nabla_k v^j) \dot{x}^k\},$$

which, by Eq. (2.8b), is same as on the RHS of Eq. (2.19). //

§ 3. Lie derivative of connection parameters

The connection parameters Γ^{*i}_{jk} (of Cartan) satisfy the coordinate transformation law

$$\tilde{\Gamma}^{*i}_{jk} = (\partial_a \bar{x}^i)\left\{\Gamma^{*a}_{bc}(\bar{\partial}_j x^b)(\bar{\partial}_k x^c) + \bar{\partial}_j \bar{\partial}_k x^a\right\}, \quad (3.1)$$

where $\bar{\partial}_j$ is given by Eq. (5.1.19). Putting from Eqs. (5.4.2), (5.4.6) and (5.4.7), the RHS of Eq. (3.1) becomes

$$\Gamma^{*i}_{jk} - \varepsilon\{\partial_j \partial_k v^i - \Gamma^{*a}_{jk}\, \partial_a v^i + \Gamma^{*i}_{ak} \partial_j v^a + \Gamma^{*i}_{ja} \partial_k v^a\}. \quad (3.2)$$

Hence, the variation $d^m \Gamma^{*i}_{jk}$, by Eq. (2.2), is

$$d^m \Gamma^{*i}_{jk} \equiv \tilde{\Gamma}^{*i}_{jk}(\bar{x}^k, \dot{\bar{x}}^k) - \Gamma^{*i}_{jk}(x^k, \dot{x}^k)$$

$$= -\varepsilon\{\partial_j \partial_k v^i - \Gamma^{*a}_{jk}\, \partial_a v^i + \Gamma^{*i}_{ak} \partial_j v^a + \Gamma^{*i}_{ja} \partial_k v^a\}, \quad (3.3)$$

whereas the change $d^v \Gamma^{*i}_{jk}$ can be obtained from Eq. (2.1). Hence, the formula (2.3) reads as

$$£\Gamma^{*i}_{jk} = \partial_j \partial_k v^i - \Gamma^{*a}_{jk}\, \partial_a v^i + \Gamma^{*i}_{ak} \partial_j v^a + \Gamma^{*i}_{ja} \partial_k v^a$$

$$+ (\partial_a \Gamma^{*i}_{jk}) v^a + (\dot{\partial}_a \Gamma^{*i}_{jk})(\partial_b v^a)\, \dot{x}^b. \quad (3.4)$$

Evaluating the second order covariant derivative of the vector v^i by means of the formulae (4.5.14) and (5.5.5a):

$$\nabla_j(\nabla_k v^i) = \partial_j (\nabla_k v^i) - (\dot{\partial}_a \nabla_k v^i) G^a_j + (\nabla_k v^a)\Gamma^{*i}_{ja} - (\nabla_a v^i)\Gamma^{*a}_{jk}$$

$$= \partial_j \partial_k v^i + (\partial_j v^a)\Gamma^{*i}_{ak} + v^a (\partial_j \Gamma^{*i}_{ak}) - v^b (\dot{\partial}_a \Gamma^{*i}_{bk}) G^a_j$$

$$+ (\nabla_k v^a)\Gamma^{*i}_{ja} - (\nabla_a v^i)\Gamma^{*a}_{jk}. \quad (3.5)$$

Eliminating $\partial_j \partial_k v^i$ from Eqs. (3.4) by means Eq. (3.5), and arranging the like terms, there follows after simplification:

$$£\Gamma^{*i}_{jk} = \nabla_j \nabla_k v^i + v^a \{\partial_a \Gamma^{*i}_{jk} - \partial_j \Gamma^{*i}_{ak} + (\dot{\partial}_b \Gamma^{*i}_{ak}) G^b_j\}$$

$$+(\nabla_a v^i - \partial_a v^i)\Gamma^{*a}_{jk} + (\dot{\partial}_a \Gamma^{*i}_{jk})(\nabla_b v^a - v^c \Gamma^{*a}_{cb})\dot{x}^b$$

$$+ \ (\partial_k v^a - \nabla_k v^a)\Gamma^{*i}_{aj} = \nabla_j \nabla_k v^i + (\dot{\partial}_a \Gamma^{*i}_{jk})(\nabla_b v^a)\dot{x}^b$$

$$+v^a \{\partial_a \Gamma^{*i}_{jk} - (\dot{\partial}_c \Gamma^{*i}_{jk})G^c_a + \Gamma^{*b}_{jk}\Gamma^{*i}_{ab}$$

$$-\partial_j \Gamma^{*i}_{ak} + (\dot{\partial}_b \Gamma^{*i}_{ak})G^b_j - \Gamma^{*c}_{ak}\Gamma^{*i}_{cj}\}$$

$$= \ \nabla_j \nabla_k v^i + v^a K^i_{ajk} + (\dot{\partial}_a \Gamma^{*i}_{jk})(\nabla_b v^a)\dot{x}^b, \qquad (3.6)$$

by Eq. (4.7.10).

Note 3.1. Employing Berwald's covariant derivation and his corresponding curvature, the Lie derivative of Berwald's connection parameters can also be similarly derived [107], Eq.(2.1), [189], p. 220, [234], p. 168:

$$£G^i_{jk} = v^a H^i_{ajk} + \mathfrak{B}_j \mathfrak{B}_k v^i + (\dot{\partial}_a G^i_{jk})(\mathfrak{B}_h v^a)\dot{x}^h. \qquad (3.7)$$

Its repeated transvections with $\dot{x}^i$'s, for Eqs. (4.6.2a) and (4.6.3), (4.9.15b), (4.9.16) and (4.10.15), also yields

$$£G^i_j = v^a H^i_{aj} + (\mathfrak{B}_j \mathfrak{B}_k v^i)\dot{x}^k. \qquad (3.8)$$

and

$$2£G^i = v^a H^i_a + (\mathfrak{B}_j \mathfrak{B}_k v^i)\dot{x}^j \dot{x}^k. \qquad (3.9)$$

Also, employing Rund's covariant derivation and the corresponding curvature tensor $\widetilde{K}^i_{ajk}$ the Lie derivative of Rund's connection parameters P^{*i}_{jk} is derived [107], Eq. (2.7):

$$£P^{*i}_{jk} = \mathfrak{R}_j \mathfrak{R}_k v^i + v^a \widetilde{K}^i_{ajk} + (\dot{\partial}_a P^{*i}_{jk})(\mathfrak{R}_h v^a)\dot{x}^h. \qquad (3.10)$$

§ 4. Lie derivative of projective connection parameters

The coordinate transformation law of the projective connection parameters Π^i_{jk} is given by Eq. (5.1.18). Putting from Eqs. (5.4.2), (5.4.6) and (5.4.7), the same reduces to

$$\widetilde{\Pi}^{i}_{jk} = (\delta^{i}_{a} + \varepsilon\,\partial_{a} v^{i})\left\{\Pi^{a}_{bc}(\delta^{b}_{j} - \varepsilon\,\partial_{j} v^{b})(\delta^{c}_{k} - \varepsilon\,\partial_{k} v^{c}) - \varepsilon\,\partial_{j}\,\partial_{k} v^{a}\right\}$$

$$= \Pi^{i}_{jk} - \varepsilon\{\partial_{j}\partial_{k}\, v^{i} - \Pi^{a}_{jk}\,\partial_{a}\, v^{i} + \Pi^{i}_{ak}\,\partial_{j}\, v^{a} + \Pi^{i}_{ja}\,\partial_{k}\, v^{a}\}$$

$$+\varepsilon\left\{(\delta^{i}_{j}\,\partial_{k} + \delta^{i}_{k}\partial_{j})\,\partial_{a}\, v^{a})\right\}/(n+1). \qquad (4.1)$$

Hence the variation $d^{m}\,\Pi^{i}_{jk}$, by Eq. (2.2), is

$$d^{m}\,\Pi^{i}_{jk} \equiv \widetilde{\Pi}^{i}_{jk}\,(\overline{x}^{k},\dot{\overline{x}}^{k}) - \Pi^{i}_{jk}(x^{k},\dot{x}^{k})$$

$$= -\varepsilon\{\partial_{j}\partial_{k}\, v^{i} - \Pi^{a}_{jk}\,\partial_{a}\, v^{i} + \Pi^{i}_{ak}\,\partial_{j}\, v^{a} + \Pi^{i}_{ja}\,\partial_{k}\, v^{a}\}$$

$$+\varepsilon\,(\delta^{i}_{j}\,\partial_{k} + \delta^{i}_{k}\partial_{j})(\partial_{a}\, v^{a})/(n+1). \qquad (4.2)$$

whereas the change $d^{v}\,\Pi^{i}_{jk}$ is obtainable from Eq. (2.1). Hence, the formula (2.3a) reads as

$$\pounds\Pi^{i}_{jk} = \partial_{j}\partial_{k}\, v^{i} - \Pi^{a}_{jk}\,\partial_{a}\, v^{i} + \Pi^{i}_{ak}\,\partial_{j}\, v^{a} + \Pi^{i}_{ja}\,\partial_{k}\, v^{a}\}$$

$$-(\delta^{i}_{j}\,\partial_{k} + \delta^{i}_{k}\partial_{j})(\partial_{a}\, v^{a})/(n+1)$$

$$+\ (\partial_{a}\,\Pi^{i}_{jk})\, v^{a} + (\dot{\partial}_{a}\,\Pi^{i}_{jk})(\partial_{b} v^{a})\,\dot{x}^{b}. \qquad (4.3a)$$

Employing the process of projective covariant derivation defined by Eq. (5.1.20) and proceeding similarly as in the previous Section, after simplification, above formula is transformed as [131], Eq. (3.5):

$$\pounds\Pi^{i}_{jk} = \mathcal{P}_{j}\,\mathcal{P}_{k}\, v^{i} + v^{a} Q^{i}_{a\,jk} + (\dot{\partial}_{a}\,\Pi^{i}_{jk})(\mathcal{P}_{b}\, v^{a})\,\dot{x}^{b}, \quad (4.3b)$$

where $Q^{i}_{a\,jk}$ are the curvature like quantities given by Eq. (5.1.24).

§ 5. Lie derivative of the projective quantities Q^{i}_{ikh}

The coordinate transformation law for these projective quantities may be quite cumbersome, we adopt a different method. Let the process of Lie derivation deform the space F_n onto $\overline{F}_n$ so that a geometric objet, say $\Omega\ (x^{k},\dot{x}^{k})$ of F_n deforming onto $\overline{\Omega}\,(x^{k},\dot{x}^{k})$ of $\overline{F}_n$ is given by [234], p. 188 and [104], Eq. (2.5):

$$\overline{\Omega}(x^k, \dot{x}^k) = \Omega(x^k, \dot{x}^k) + \varepsilon \pounds \Omega(x^k, \dot{x}^k). \quad (5.1)$$

Thus, let the projective quantities connection Q^{i}_{ikh} deform onto $\overline{Q}^{i}_{ikh}$. Writing their intrinsic value in terms of the deformed projective connection parameters $\overline{\Pi}^{i}_{jk}$ in analogy with Eq. (5.1.24):

$$\overline{Q}^{i}_{jkh} = 2\{\partial_{[j}\overline{\Pi}^{i}_{k]h} + (\dot{\partial}_a \overline{\Pi}^{i}_{h[j})\overline{\Pi}^{a}_{k]} + \overline{\Pi}^{i}_{a[j})\overline{\Pi}^{a}_{k]h}\}. \quad (5.2)$$

Substituting for the deformed values of the connection parameters obtainable from Eq. (5.1):

$$\overline{\Pi}^{i}_{jk}(x^k, \dot{x}^k) = \Pi^{i}_{jk}(x^k, \dot{x}^k) + \varepsilon \pounds \Pi^{i}_{jk}(x^k, \dot{x}^k), \quad (5.3)$$

together with

$$\overline{\Pi}^{i}_{j}(x^k, \dot{x}^k) = \Pi^{i}_{j}(x^k, \dot{x}^k) + \varepsilon \pounds \Pi^{i}_{j}(x^k, \dot{x}^k), \quad (5.4)$$

in above equation and simplifying by means of Eqs. (5.1.24), we derive

$$\pounds Q^{i}_{jkh} = \lim_{\varepsilon \to 0} (\overline{Q}^{i}_{jkh} - Q^{i}_{jkh})/\varepsilon$$

$$= \mathcal{P}_j(\pounds \Pi^{i}_{kh}) - \mathcal{P}_k(\pounds \Pi^{i}_{jh}) - (\dot{\partial}_a \Pi^{i}_{kh})(\pounds \Pi^{a}_{j}) + (\dot{\partial}_a \Pi^{i}_{jh}(\pounds \Pi^{a}_{k}). \quad (5.5)$$

§ 6. Commutation formulae involving Lie derivation

Theorem 6.1. Lie derivative commutes with the partial derivation w.r.t. directional coordinates:

$$(\pounds \dot{\partial}_k - \dot{\partial}_k \pounds) X^i = 0, \quad (6.1)$$

for any arbitrary contravariant vector field X^i.

Proof. The derivative $\dot{\partial}_k X^i$ being a mixed tensor, its Lie derivative can be found by Eq. (2.16a):

$$\pounds \dot{\partial}_k X^i = v^a \nabla_a \dot{\partial}_k X^i - (\dot{\partial}_k X^a) \nabla_a v^i + (\dot{\partial}_a X^i) \nabla_k v^a$$

$$+ (\dot{\partial}_a \dot{\partial}_k X^i)(\nabla_b v^a) \dot{x}^b. \quad (6.2)$$

On the other hand, for the independence of the generator v^i of the infinitesimal transformation on the directional coordinates and vectorial character of Lie derivative of a vector given by Eq. (2.8b), its partial derivative w.r.t. $\dot{x}^j$ is

$$\dot{\partial}_k \pounds X^i = v^a \dot{\partial}_k \nabla_a X^i - (\dot{\partial}_k X^a) \nabla_a v^i$$

$$- X^a \dot{\partial}_k \nabla_a v^i + (\dot{\partial}_k \dot{\partial}_a X^i)(\nabla_b v^a) \dot{x}^b. \qquad (6.3)$$

Subtraction of Eq. (6.3) from Eq. (6.2), for Eqs. (4.6.4), (4.8.2) and cancellation of certain common terms yields the desired result. //

Theorem 6.2. Lie derivative commutes with the covariant derivation process of Cartan according to

$$(\pounds \nabla_k - \nabla_k \pounds) X^i = X^j \pounds \Gamma^{*i}_{jk} - (\dot{\partial}_j X^i) \pounds G^j_k, \qquad (6.4)$$

for any arbitrary contravariant vector field X^i.

Proof. The derivative $\nabla_k X^i$ being a mixed tensor, its Lie derivative can be found by Eq. (2.15):

$$\pounds \nabla_k X^i = v^a \nabla_a \nabla_k X^i - (\nabla_k X^a) \nabla_a v^i + (\nabla_a X^i) \nabla_k v^a$$

$$+ (\dot{\partial}_a \nabla_k X^i)(\nabla_b v^a) \dot{x}^b. \qquad (6.5)$$

On the other hand, for vectorial character of Lie derivative of a vector given by Eq. (2.8b), its covariant derivative w.r.t. x^k, by Eqs. (4.5.14) and (4.5.16) is

$$\nabla_k \pounds X^i = (\nabla_k v^a) \nabla_a X^i + v^a \nabla_k \nabla_a X^i - (\nabla_k X^a)(\nabla_a v^i) - X^a (\nabla_k \nabla_a) v^i$$

$$+ \{(\nabla_k \dot{\partial}_a X^a)(\nabla_b v^a) + (\dot{\partial}_a X^a)(\nabla_k \nabla_b v^a)\} \dot{x}^b. \qquad (6.6)$$

Subtraction of Eq. (6.6) from Eq. (6.5), cancelling certain common terms and re-arranging the like terms, we get

$$(\pounds \nabla_k - \nabla_k \pounds) X^i = 2 v^a \nabla_{[a} \nabla_{k]} X^i + \{(\dot{\partial}_a \nabla_k - \nabla_k \dot{\partial}_a) X^i\} (\nabla_b v^a) \dot{x}^b$$

$$+ X^a (\nabla_k \nabla_a v^i) - (\dot{\partial}_a X^i)(\nabla_k \nabla_b v^a) \dot{x}^b$$

$$= v^a \{ K^i_{akh} X^h - (\dot{\partial}_j X^i) K^j_{akh} \dot{x}^h \}$$

$$+\{(\dot{\partial}_a \Gamma^{*i}_{kh}) X^h - (\nabla C^j_{ak})(\dot{\partial}_j X^i)\}(\nabla_b v^a)\dot{x}^b$$

$$+ X^a(\nabla_k \nabla_a v^i) - (\dot{\partial}_a X^i)(\nabla_k \nabla_b v^a)\dot{x}^b,$$

by Eqs. (4.6.4), (4.7.11) and (4.8.2). Collecting the coefficients of X^h and $\dot{\partial}_j X^i$ and re-arranging the terms, above equation reduces to

$$(£\nabla_k - \nabla_k £) X^i = X^h \{\nabla_k \nabla_h v^i + v^a K^i_{akh} + (\dot{\partial}_a \Gamma^{*i}_{kh})(\nabla_b v^a)\dot{x}^b$$

$$-(\dot{\partial}_j X^i)\{\nabla_k \nabla_h v^j + v^a K^j_{akh} + (\dot{\partial}_a \Gamma^{*j}_{kh})(\nabla_b v^a)\dot{x}^b\}\dot{x}^h,$$

which, for Eqs. (4.6.2b), (2.9b) and (3.6), assumes the desired form. //

Note 6.1. Analogous to Eq. (6.4), the commutation rule for the Lie derivation with the operators of Berwald's covariant derivation and projective covariant derivation are also derived by the author [107], [131]:

$$(£ \mathfrak{B}_k - \mathfrak{B}_k £) X^i = X^j £ G^i_{jk} - (\dot{\partial}_j X^i) £ G^j_k, \tag{6.7}$$

and

$$(£ \mathcal{P}_k - \mathcal{P}_k £) X^i = X^j £ \Pi^i_{jk} - (\dot{\partial}_j X^i) £ \Pi^j_k, \tag{6.8}$$

where Π^j_k is given by Eq. (5.1.14) for any arbitrary contravariant vector field X^i.

Theorem 6.3. Lie derivative commutes with Cartan's first process of covariant derivation according to

$$(£\dot{\nabla}_k - \dot{\nabla}_k £) X^i = X^j £ C^i_{jk}, \tag{6.9}$$

for any arbitrary contravariant vector field X^i.

Proof. The derivative $\dot{\nabla}_k X^i$ given by Eq. (4.5.13) is a mixed tensor, so its Lie derivative can be found by Eq. (2.15):

$$£\dot{\nabla}_k X^i = £(\dot{\partial}_k X^i + X^j C^i_{jk}) = £\dot{\partial}_k X^i + (£X^j) C^i_{jk} + X^j £C^i_{jk}. \tag{6.10}$$

On the other hand, for vectorial character of Lie derivative of a vector given by Eq. (2.8b), its Cartan's first covariant derivative, by Eq. (4.5.13), is

$$\dot{\nabla}_k £ X^i = \dot{\partial}_k (£ X^i) + (£X^j) C^i_{jk}. \tag{6.11}$$

Subtracting Eq. (6.11) from Eq. (6.10), cancelling common terms and using Eq. (6.1), we get the desired result. //

Note 6.2. The formula in Eq. (6.1) also holds for the connection parameters and projective connection parameters [107]:

$$(\pounds \dot{\partial}_j - \dot{\partial}_j \pounds) G^i_{kh} = (\pounds \dot{\partial}_j - \dot{\partial}_j \pounds) \Pi^i_{kh} = 0. \qquad (6.12)$$

§ 7. Motion

The distance preserving infinitesimal transformations have been called motion (cf. Chapter 5, § 5). Necessary and sufficient conditions for the same are derived vide Eq. (5.5.4) alternately Eq. (5.5.8).

Applying the formula (2.15) and noting Eq. (4.5.15), the Lie derivative of the metric tensor can be deduced:

$$\begin{aligned} \pounds g_{ij} &= g_{aj} \nabla_i v^a + g_{ia} \nabla_j v^a + (\dot{\partial}_k g_{ij})(\nabla_h v^k) \dot{x}^h \\ &= \nabla_i v_j + \nabla_j v_i + 2 C_{ijk})(\nabla_h v^k) \dot{x}^h, \end{aligned} \qquad (7.1)$$

by Eqs. (3.9.1), (4.5.15) and (5.5.6). It may be noted that above expression vanishes by Eq. (5.5.8):

$$\pounds g_{ij} = 0 \qquad (7.2)$$

for a motion. Hence, we have the

Theorem 7.1. The infinitesimal transformation vide Eq. (5.4.1) defines a motion iff the Lie derivative of the metric tensor vanishes, i.e. there hold Killing equations.

Corollary 7.1. The tensors g^{ij}, C_{ijk} and C^i_{jk} are Lie-invariants in the space admitting a motion.

Proof. (i) Taking Lie derivative of Eq. (4.2.1) and using linear property of the operator £ and Eq. (7.2), there also results

$$g_{ij} \pounds g^{jk} = 0.$$

The same on transvection with g^{hi}, for Eq. (4.2.1), yields

$$\pounds g^{hk} = 0 . \tag{7.3}$$

(ii) The Lie derivation of Eq. (3.9.1), for commutation rule given by Eq. (6.1) and Eq. (7.2), proves the second result:

$$\pounds C_{ijk} = (1/2) \pounds \dot{\partial}_i g_{jk} = (1/2)\, \dot{\partial}_i \pounds g_{jk} = 0. \tag{7.4}$$

(iii) Transvection of above equation by g^{hj}, for Eqs. (4.2.11) and (7.3), yields the third result:

$$g^{hj} \pounds C_{ijk} = \pounds (g^{hj} C_{ijk}) = \pounds C^h_{ik} = 0. \tag{7.5}$$

Theorem 7.2. The integrability conditions for Eq. (7.2) are

$$\pounds \Gamma^{*i}_{jk} = 0. \tag{7.6}$$

Proof. Applying the commutation formula (6.4) to the metric tensor and putting from Eqs. (3.9.1) and (4.5.15), there follows

$$\nabla_k \pounds g_{ij} = g_{aj} \pounds \Gamma^{*a}_{ik} + g_{ia} \pounds \Gamma^{*a}_{jk} + 2C_{aij} \pounds G^a_k . \tag{7.7}$$

Interchanging the indices *i*, *j*, *k* cyclically we get two more such equations. Subtracting the third one from the sum of the first two Eqs. There follows

$$2g_{aj} \pounds \Gamma^{*a}_{ik} = \nabla_k \pounds g_{ij} + \nabla_i \pounds g_{jk} - \nabla_j \pounds g_{ki}$$

$$- 2(C_{aij} \pounds G^a_k + C_{ajk} \pounds G^a_i - C_{aki} \pounds G^a_j).$$

Its transvection by g^{jh}, for Eqs. (4.2.1) and (4.2.11), yields

$$2\pounds \Gamma^{*h}_{ik} = g^{jh} (\nabla_k \pounds g_{ij} + \nabla_i \pounds g_{jk} - \nabla_j \pounds g_{ki})$$

$$- 2(C^h_{ai} \pounds G^a_k + C^h_{ak} \pounds G^a_i - g^{jh} C_{aki} \pounds G^a_j),$$

which, for Eq. (7.2) holding in a space admitting a motion, reduces to

$$\pounds \Gamma^{*h}_{ik} = -(C^h_{ai} \pounds G^a_k + C^h_{ak} \pounds G^a_i - g^{jh} C_{aki} \pounds G^a_j). \tag{7.8}$$

Its transvection with $\dot{x}^k$, for Eqs. (4.2.9), (4.2.14), (4.6.2b), (4.6.3) and (2.9b), also gives

$$\pounds\, G_i^h = -2C_{a\,i}^h\, \pounds\, G^a\ . \tag{7.9}$$

Finally, its transvection with $\dot{x}^i$, for Eqs. (4.2.14), (4.6.3) and (2.9b), determines

$$\pounds\, G^h = 0, \tag{7.10}$$

in a space admitting a motion. Reverting back to above discussion, Eq. (7.10) causes vanishing of $\pounds\, G_i^h$ from Eq. (7.9) and subsequent vanishing of $\pounds\, \Gamma_{i\,k}^{*h}$ from Eq. (7.8). //

Note 7.1. Thus, in a Finsler space admitting a motion the Lie derivatives of the functions G^i, G_j^i and the connection parameters $\Gamma_{j\,k}^{*\,i}$ vanish identically:

$$\pounds\, \Gamma_{j\,k}^{*\,i} = \pounds\, G_j^i = \pounds\, G^i = 0. \tag{7.11}$$

Theorem 7.3. The integrability conditions for Eq. (7.2) are alternately given by

$$\pounds\, G_{jk}^i = 0. \tag{7.12}$$

Proof. Employing the commutation formula (6.7) and proceeding similarly as in the previous theorem, analogue of Eq. (7.8), gets an additional term $3\pounds\nabla C_{i\,jk}$:

$$\pounds\, G_{ik}^h = -(C_{a\,i}^h\, \pounds\, G_k^a + C_{a\,k}^h\, \pounds\, G_i^a - g^{j\,h}\, C_{a\,k\,i} \pounds\, G_j^a) + 3\pounds\nabla C_{i\,jk}. \tag{7.13}$$

Transvection of above equation with $\dot{x}^k$ reduces the additional term to zero and the rest of the treatment is same as in the previous theorem. //

Corollary 7.2. Operations of covariant and Lie derivation commute in F_n admitting motion.

Proof. The result follows immediately from Eqs. (6.7), (7.11) and (7.12). //

§ 8. Affine motion

Transformations preserving parallelism of vectors are called *affine motions*. Necessary and sufficient conditions for the same require vanishing of Lie derivative of connection parameters [234]. Thus, from Eqs. (3.7) - (3.9), there result the conclusions:

$$\pounds G^i_{jk} = \mathfrak{B}_j \mathfrak{B}_k v^i + v^a H^i_{ajk} + (\dot{\partial}_a G^i_{jk})(\mathfrak{B}_h v^a)\dot{x}^h = 0, \quad (8.1)$$

$$\pounds G^i_j = v^a H^i_{aj} + (\mathfrak{B}_j \mathfrak{B}_k v^i)\dot{x}^k = 0, \quad (8.2)$$

and

$$2\pounds G^i = v^a H^i_a + (\mathfrak{B}_j \mathfrak{B}_k v^i)\dot{x}^j \dot{x}^k = 0. \quad (8.3)$$

in the space F_n admitting an affine motion.

Theorem 8.1. The functions G^i, G^i_j, G^i_{ij} and the tensor G^i_{ijk} also have the vanishing Lie derivatives in a space admitting an affine motion, i.e. they experience no change under the said transformation:

$$\pounds G^i = \pounds G^i_j = \pounds G^i_{ij} = \pounds G^i_{jkh} = 0, \quad (8.4)$$

Proof. Transvections and derivation of Eq. (8.1) - (8.3) w.r.t. $\dot{x}^i$'s and contraction of indices *i*, *k* , for Eqs. (4.6.2a), (4.6.3), (4.9.12) and (2.9b) establish the results. //

Theorem 8.2. Every motion is an affine motion.

Proof. The proof is a direct consequence of Theorems 7.1 and 7.2 (alternately 7.3). //

Note 8.1. Obviously the Corollary 7.2 also holds in F_n admitting an affine motion.

Theorem 8.3. The integrability conditions for Eq. (8.1) are

$$\pounds H^i_{jkh} = 0 \quad (8.5); \qquad \pounds \dot{\partial}_l H^i_{jkh} = 0; \quad (8.6)$$

and

$$\pounds \mathfrak{B}_{l_1} H^i_{jkh} = \pounds \mathfrak{B}_{l_1} \mathfrak{B}_{l_2} H^i_{jkh} = \pounds \mathfrak{B}_{l_1} \mathfrak{B}_{l_2} \ldots \mathfrak{B}_{l_p} H^i_{jkh} = 0. \quad (8.7)$$

Proof. (i) Let the curvature tensor (of Berwald) H^i_{jkh} deform under the process of Lie derivation according to Eq. (5.1). Writing the intrinsic expression for deformed curvature tensor in terms of the deformed connection parameters in analogy with Eq. (4.7.15):

$$\overline{H}^i_{jkh} = 2\{\partial_{[j}\overline{G}^i_{k]h} + (\dot{\partial}_a \overline{G}^i_{h[j})\overline{G}^a_{k]} + \overline{G}^i_{a[j})\overline{G}^a_{k]h}\}, \quad (8.8)$$

and proceeding on the lines similar to those in § 5, in above equation and simplifying by means of Eqs. (4.7.15), we derive

$$£H^i_{jkh} = \lim_{\varepsilon \to 0} (\overline{H}^i_{jkh} - H^i_{jkh}) / \varepsilon$$

$$= \mathfrak{B}_j (£G^i_{kh}) - \mathfrak{B}_k (£G^i_{jh}) - G^i_{akh}(£G^a_j) + G^i_{ajh}(£G^a_k). \quad (8.9)$$

Thus, for an affine motion, the Eq. (7.12) and therefore Eq. (7.11), necessarily cause vanishing of Lie derivative of the curvature tensor.

(ii) The directional derivative of Eq. (8.5), for Eq. (6.1), yields the result in Eq. (8.6).

(iii) Next, applying the commutation rule given by Eq. (6.7) to the curvature tensor H^i_{ikh} and putting from Eqs. (8.1), (8.4) and (8.5) we get the first result in Eq. (8.7). Forming successive covariant derivations of this result and applying the commutation formula (6.7) repeatedly we also derive further results in Eq. (8.7). //

Corollary 8.1. The Lie derivatives of the Berwald's deviation tensor and other associated tensors also vanish in a space admitting an affine motion.

$$£H^i_{jk} = £H^i_j = £H_{kh} = £H_k = £H = 0. \quad (8.10)$$

Proof. Repeated transvections and contractions of Eq. (8.5) and use of Eqs. (4.9.15b), (4.9.16), (4.9.21) and (2.9b) establish the results. //

In addition, there also hold the following results as immediate consequences of Eqs. (6.1) and (8.10) in a space admitting an affine motion:

$$£\dot{\partial}_l H_{kh} = £\dot{\partial}_l H_k = £\dot{\partial}_l H = 0. \quad (8.11)$$

Note 8.2. In view of Theorem 8.2, the integrability conditions in above Theorem along with Eqs. (8.10) and (8.11) also hold good in a space admitting a motion.

Corollary 8.2. The projective tensors W^i_j, W^i_{jk} and W^i_{jkh} are Lie-invariants in a space admitting an affine motion:

$$£W^i_j = £W^i_{jk} = £W^i_{jkh} = 0. \quad (8.12)$$

Proof. Putting from Eqs. (6.2.9b), (6.8.5) and (6.8.10) in Eq. (5.2.13c) one immediately gets the second of Eqs. (8.12); while its transvection by $\dot{x}^k$, for Eqs. (6.2.9b) and (5.2.17), yields the first one. On the other hand, the directional derivative of the second equation w.r.t. $\dot{x}^h$, for commutation rule vide Eq. (6.1) and Eq. (5.2.14) yields the third result. //

8.1. Implications of Eq. (8.1)

The Eqs. (8.1) - (8.3) express necessary and sufficient conditions for an affine motion in a F_n. The transvection of Eq. (8.2) with vector v^j, for the skew-symmetric properties of the tensor H^i_{aj} makes the first term zero:

$$v^j v^a H^i_{aj} = 0, \tag{8.13}$$

leaving the transvected equation as

$$v^j (\mathfrak{B}_j \mathfrak{B}_k v^i)\dot{x}^k = 0. \tag{8.14}$$

On the other hand, contraction of Eq. (8.2) for the indices *i* and *j*, gives

$$v^a H^i_{ai} + (\mathfrak{B}_j \mathfrak{B}_k v^j)\dot{x}^k = 0. \tag{8.15}$$

Also, a covariant derivation of Eq. (8.2) w.r.t. x^h, for Eq. (4.6.7), gives

$$(\mathfrak{B}_h v^a) H^i_{aj} + v^a \mathfrak{B}_h H^i_{aj} + \mathfrak{B}_h (\mathfrak{B}_j \mathfrak{B}_k v^i \dot{x}^k) = 0, \tag{8.16}$$

8.2. Affine motions generated by special vector fields

Different cases of vector field generating an infinitesimal transformation have been considered vide Eqs. (5.9.1) and (5.9.3) - (5.9.8). Accordingly, for these respective cases, Eqs. (8.14) and (8.15) reduce to:

Vector field	Eq. (8.14)	Eq. (8.15)
Contra and concurrent	Identically zero	$v^a H^i_{ai} = 0$ (8.15a)
Special concircular	$v^j \rho_j \dot{x}^i = 0$ $\Rightarrow$ $v^j \rho_j = 0$, (8.14a) (for arbitrary $\dot{x}^i$)	$v^a H^i_{ai} + \rho_k \dot{x}^k = 0$, (8.15b)

Recurrent	$v^i(\mu_{jk} v^j \dot{x}^k) = 0$ $\Rightarrow \mu_{jk} v^j \dot{x}^k = 0$, (8.14b) (for arbitrary v^i)	$v^a H^i_{ai} + \mu_{jk} v^j \dot{x}^k = 0$ $\Rightarrow v^a H^i_{ai} = 0$, (8.15c) by Eq. (8.14b)
Concircular	$v^j \rho_j \dot{x}^i + v^i(\mu_{jk} v^j \dot{x}^k) = 0.$ (8.14c)	$v^a H^i_{ai} + (\rho_k + \mu_{jk} v^j)\dot{x}^k = 0.$ (8.15d)

8.3. Affine motions generated by special concircular vector fields

As per assumption, the vector field v^i is independent of the directional arguments, its covariant derivative $\mathfrak{B}_h v^i$, found by Eq. (4.6.6), is positively homogeneous of degree zero in $\dot{x}^i$'s, and so is also the function ρ in Eq. (5.9.1c). Therefore, by Euler's theorem of calculus, we have

$$\dot{x}^i \dot{\partial}_i \rho = 0. \tag{8.17}$$

It is seen vide Eq. (8.14a) that the vector field v^i generating a special concircular affine motion is orthogonal to the gradient vector field ρ_j defined by Eq. (5.9.4). Thus, the Theo. 4.1 of authors' work [135] holds in a general Finsler space.

Further, the Lie derivative of the scalar function ρ may be found by Eq. (2.3d). The same, for Eqs. (5.9.1c) and (8.14a) reduces to

$$\pounds \rho = \rho\, \dot{x}^j (\dot{\partial}_j \rho) = 0, \tag{8.18}$$

for Eq. (8.17). Thus, the function ρ is a Lie-invariant in the space F_n admitting a special concircular affine motion.

§ 9. Projective motion

Transformations mapping geodesics of F_n over geodesics of $\overline{F}_n$ have been called *projective* transformations. It is seen (cf. Chapter 5, § 1) that the connection parameters of the two spaces preserving geodesics are connected by Eq. (5.1.7) for some scalar functions p with their partial derivatives p_j and p_{jk} defined by Eqs. (5.1.9). Accordingly, the infinitesimal point change given by Eq. (5.4.1) gives rise to a *projective motion* if the change $\varepsilon\,£\,G^i_{jk}$ in the connection parameters coincides with the change given by Eq. (5.1.7):

$$£\,G^i_{jk} = 2\delta^i_{(j}\, p_{k)} + \dot{x}^i\, p_{jk}\,. \tag{9.1}$$

The corresponding changes in the functions $G^i, G^i_j,\ G^i_{ij}$ and G^i_{ijk}, as given by Eqs. (5.1.5), (5.1.6), (5.1.11) and (5.1.12), assume the forms

$$\left.\begin{aligned} £\,G^i &= \dot{x}^i\, p, & £\,G^i_j &= \delta^i_j\, p + \dot{x}^i\, p_j\,, \\ £\,G^i_{ij} &= (n+1)\, p_j\,, & £\,G^i_{ijk} &= (n+1)\, p_{jk}\,. \end{aligned}\right\} \tag{9.2}$$

Accordingly, the projective connection parameters given by Eq. (5.1.13) assume the change

$$£\,\Pi^i_{jk} = £\,G^i_{jk} - \{2\delta^i_{(j}\, £\,G^r_{k)r} + \dot{x}^i\, £\,G^r_{jkr}\}/(n+1) = 0, \tag{9.3}$$

by Eqs. (9.1) and (9.2). Thus, we have the

Theorem 9.1. The necessary and sufficient conditions for infinitesimal transformations to define a projective motion require vanishing of Lie derivative of projective connection parameters [234].

Corollary 9.1. There also hold the following results in a space admitting a projective motion:

$$£\,\Pi^i_j = £\,\Pi^i_{jkh} = 0. \tag{9.4}$$

Proof. Transvection of Eq. (9.3) by $\dot{x}^k$ and derivation of the same w.r.t. $\dot{x}^h$, for Eqs. (5.1.14), (5.1.15) and (6.12) respectively prove the results.//

Corollary 9.2. Every affine motion is a projective motion.

Proof. Eqs. (6.12), (9.1) and (9.3) prove the result. //

Theorem 9.2. The integrability conditions for Eq. (9.3) are

$$£\,Q^i_{jkh} = 0 \quad (9.5); \quad \text{and} \quad £\,W^i_{jkh} = 0. \tag{9.6}$$

Proof. The first result is a direct consequence of Eqs. (5.5), (9.3) and (9.4).

Next, contraction of Eq. (9.5), for Eqs. (5.1.26), also yields

$$\pounds Q_{kh} = 0, \tag{9.7}$$

in the space admitting a projective motion. Thus, for Eqs. (9.5) and (9.7), the Eq. (5.2.27) establishes the second result. //

Corollary 9.3. As necessary consequences of Eq. (9.6), there also hold the results

$$\pounds W^i_{jk} = 0 \quad (9.8); \quad \text{and} \quad \pounds W^i_j = 0. \tag{9.9}$$

Proof. Successive transvections of Eq. (9.6) with $\dot{x}^h$ and $\dot{x}^k$, for Eqs. (5.2.16), (5.2.17) and (2.9b), immediately yield the desired results. //

9.1. Special concircular projective motion

Along with the discussion of Sub-section 9.1 of the previous chapter if there also holds Eq. (9.1), the infinitesimal transformation given by Eq. (5.4.1) will be said to define a special concircular projective motion in an F_n. Thus, from Eqs. (4.10.15), (5.9.1c), (5.9.4), (3.7) and (9.1), there follows

$$\pounds G^i_{jk} = v^a H^i_{ajk} + \mathfrak{B}_j\, \rho\, \delta^i_k = \delta^i_j\, p_k + \delta^i_k\, p_j + \dot{x}^i\, p_{jk}. \tag{9.10}$$

On contraction w.r.t. the indices i and j, for Eqs. (4.9.20), (5.1.10b), it yields

$$\mathfrak{B}_k\, \rho - v^a H_{ak} = (n+1)\, p_k. \tag{9.11a}$$

Similarly, contraction of Eq. (9.10) w.r.t. the indices i and k, for Eqs. (4.10.3) and (5.1.10b) determines

$$n\, \mathfrak{B}_j\, \rho + v^a (H_{ja} - H_{aj}) = (n+1)\, p_j. \tag{9.11b}$$

Eliminating p_j, from Eqs. (9.11), there follows

$$(n-1)\, \mathfrak{B}_j\, \rho + v^a H_{ja} = 0. \tag{9.12}$$

As seen in Sub-section 9.3 of the previous chapter, for invariance of the function ρ under covariant derivation, ρ itself turns a constant and the transformation reduces to a contra or concurrent form. Accordingly, Eq. (9.12) reduces to Eq. (5.9.28). Also, either of Eqs. (9.11) yield

$$v^{a} H_{ak} + (n+1)\, p_{k} = 0. \tag{9.13}$$

Thus, we have the

Theorem 9.3. A special concircular projective motion with ρ as a covariant constant in an F_n is generated by either a contra or concurrent vector field and there hold the Eqs. (5.9.28) and (9.13).

Putting from Eqs. (9.1) and (9.2) in Eq. (8.9), the Lie derivative of the curvature tensor simplifies to

$$£\, H^{i}_{jkh} = 2\, \mathfrak{B}_{[j} \{ \delta^{i}_{k]}\, p_{h} + p_{k]}\, \delta^{i}_{h} + \dot{x}^{i}\, p_{k]h} \}. \tag{9.14}$$

Its contraction w.r.t. the indices *i* and *h*, for Eqs. (4.10.3) and (5.1.10b), yields

$$£\, (H_{kj} - H_{jk}) = 2\, (n+1)\, \mathfrak{B}_{[j}\, p_{k]}\,. \tag{9.15}$$

Transvection by v^{k}, for Eqs. (5.9.28), (2.9a) and (9.13), reduces its LHS as $-(n+1)\, £\, p_{j}$ and Lie derivative of p_{j} can be evaluated by Eq. (2.12c):

$$£\, p_{j} = v^{k} (\mathfrak{B}_{k}\, p_{j}) + p_{k} (\mathfrak{B}_{j}\, v^{k}) + (\dot{\partial}_{k}\, p_{j})\, (\mathfrak{B}_{h}\, v^{k})\, \dot{x}^{h}. \tag{9.16}$$

Putting from Eq. (5.9.1c) and simplifying by means of Eq. (5.1.10b), it further reduces to

$$£\, p_{j} = v^{k} (\mathfrak{B}_{k}\, p_{j}) + \rho\, p_{j}. \tag{9.17}$$

Therefore, a transvection of Eq. (9.15) by v^{k} yields

$$v^{k} (\mathfrak{B}_{j}\, p_{k}) + \rho\, p_{j} = 0. \tag{9.18a}$$

Retaining the second term as a derivative of vector field v^{i} as per Eq. (5.9.1c):

$$\rho\, p_{j} = \rho\, \delta^{k}_{j}\, p_{k} = (\mathfrak{B}_{j}\, v^{k})\, p_{k},$$

Eq. (9.18a) is interpreted as

$$v^{k} (\mathfrak{B}_{j}\, p_{k}) + (\mathfrak{B}_{j}\, v^{k})\, p_{k} = \mathfrak{B}_{j} (v^{k}\, p_{k}) = 0. \tag{9.18b}$$

Thus, we have the

Theorem 9.4. The scalar $v^k p_k$ is a covariant constant in an F_n admitting a projective motion characterized by Eqs. (5.9.1c) and (5.9.28).

Also, for Eq. (2.9a), we have

$$\pounds(v^j p_j) = v^j \pounds p_j = v^j \{ v^k (\mathfrak{B}_k p_j) + \rho p_j \}, \quad \text{by Eq. (9.17)}$$

$$= v^j \{ v^k (\mathfrak{B}_k p_j) - v^k (\mathfrak{B}_j p_k) \}, \quad \text{by Eq. (9.18a).}$$

Interchange of dummy indices in one of the above terms makes the sum zero. Hence, we have the

Theorem 9.5. The scalar $v^k p_k$ is Lie-invariant in an F_n admitting a projective motion characterized by the conditions in Eqs. (5.9.1c) and (5.9.28).

9.2. Recurrent projective motion

If a vector-field v^i satisfying Eq. (5.9.1d) defines a projective motion in an F_n, in continuation with the discussion of § 10 in Chapter 5, there also hold the Eqs. (3.8) and (9.2), which, for Eq. (5.9.5), reduce to

$$\pounds G^i_k = v^j H^i_{jk} + v^i \mu_{kh} \dot{x}^h = p \delta^i_k + \dot{x}^i p_k. \tag{9.19}$$

On transvection with v^k, for skew-symmetry of the tensor H^i_{ik} in its lower indices, above equation yields

$$v^i v^k \mu_{kh} \dot{x}^h = p v^i + w \dot{x}^i, \tag{9.20}$$

where we have put

$$w \equiv v^k p_k. \tag{9.21}$$

On the other hand, contraction of Eq. (9.19) w.r.t. the indices i and k, for Eqs. (4.9.20) and (5.1.10a), also yields

$$v^k \mu_{kh} \dot{x}^h = v^j H_j + (n+1) p. \tag{9.22}$$

Multiplying above equation by v^i and subtracting the resulting equation from Eq. (9.20), there results

$$v^i (v^j H_j + n p) = w \dot{x}^i. \tag{9.23}$$

On transvection with p_i, for Eqs. (5.1.10a) and (9.21), it finally determines the function p :

$$p = v^j H_j / (1-n). \tag{9.24}$$

Its directional derivation w.r.t. $\dot{x}^k$, for Eqs. (4.9.21b) and (5.1.9a), also yields

$$p_k = v^j H_{jk} / (1-n). \tag{9.25}$$

Thus, we have the

Theorem 9.6. The equations (9.24) and (9.25) represent the integrability conditions for a recurrent projective motion in an F_n.

§ 10. Curvature collineation

Taking vanishing Lie derivative of the Riemann curvature tensor, G.H. Katzin et al. [55] introduced the concepts of curvature collineation (onward briefly denoted by *CC*) in a Riemannian space. The concept was extended to Finsler spaces by Pande, Kumar and Khan [167], Prasad [183], Singh [204], [205], Singh and Prasad [207] and the present author with Nawal-Kishore [140]. Certain necessary consequences of a *CC* defined by vanishing of Cartan's curvature tensor K^i_{jkh} were derived in an affinely-connected Finsler space in [183]. Others [167] considered vanishing of Cartan's third curvature tensor R^i_{jkh} and studied *CC*s which are also simultaneously projective motions. *CC*s defined by vanishing of Berwald's curvature tensor were considered in [207]. All these authors considered special cases of *CC*s only when the transformation is simultaneously a projective motion and called these *CC*s as special *CC*s. Using more appropriate terminologies, the *CC*s corresponding to various types of curvature tensors in Finsler space were revisited by the present author et al. [140] and offered both trivial and non-trivial examples of such *CC*s. As a trivial case, it is seen (cf. Example 2.1 therein) that either of motion or affine motion defines a curvature collineation defined by vanishing of Lie derivative of the Berwald's curvature tensor H^i_{ikh} (hence briefly denoted by *H-CC*). Before taking up some further examples of *CC*s, deriving necessary and sufficient conditions for the same and proving some concerned results, we give a more systematic definition of different *CC*s defined in terms of various types of curvature tensors in Finsler geometry.

Definition 10.1. An infinitesimal transformation given by Eq. (5.4.1) defines a *K*-CC, *R*-CC, *H*-CC or *P*-CC if it makes the respective curvature tensors of Cartan (K^{i}_{jkh}, R^{i}_{jkh}), Berwald (H^{i}_{ikh}) and projective curvature tensor (W^{i}_{ikh}) invariant under Lie derivation, i.e. there hold the Eqs.

$$\pounds K^{i}_{jkh} = 0 \qquad (10.1); \qquad \pounds R^{i}_{jkh} = 0, \qquad (10.2)$$

(8.5) or (9.6) respectively.

10.1. *H-CC*

Thus, in view of Theorems 8.2, 8.3, and Eqs. (8.12), we have the following:

Example 10.1. Either of a motion or affine motion defines a *H-CC* as well as a *P-CC*.

Example 10.2. Every projective motion defines a *P-CC*.

The condition of affinely-connectedness imposed in [183] on the space F_n for a motion to define a *K-CC* or *R-CC* is superfluous. On contrary, the author et al. [140] prove the following more general result:

Theorem 10.1. A motion is both a *K-CC* and *R-CC* in any space F_n.

Proof. (i) For a motion in F_n, there hold the Eqs. (7.2) and (7.11). The Lie derivative of Cartan's curvature tensor K^{i}_{jkh}, derivable analogously from Eq. (8.9):

$$\pounds K^{i}_{jkh} =$$

$$\nabla_j(\pounds \Gamma^{*i}_{kh}) - \nabla_k(\pounds \Gamma^{*i}_{jh}) - (\dot{\partial}_a \Gamma^{*i}_{kh})(\pounds G^a_j) + (\dot{\partial}_a \Gamma^{*i}_{jh})(\pounds G^a_k), (10.3)$$

vanishes for Eqs. (7.11). Thus, making the motion a *K-CC* as well.

(ii) Next, taking the Lie derivative of Eq. (4.7.13), and putting from Eqs. (7.5) and (10.1), we get Eq. (10.2).

The tensors associated with Berwald's Curvature tensors are also seen Lie-invariant in the space admitting an affine motion (cf. Corollary 8.1). Simultaneously, the directional derivations of the first two of Eqs. (8.10), for the commutation rule vide Eq. (6.1) and Eqs. (4.9.15) and (4.9.18) yield back the Eq. (8.5). This establishes the:

Theorem 10.2. Vanishing Lie derivatives of Berwald's tensors H^i_{jk} and H^i_j are necessary and sufficient conditions for an infinitesimal transformation to define an *H-CC*.

10.2. *H-CC* projective motion

In this sub-section, we consider the existence of a projective motion

which is simultaneously an *H-CC* too. A projective motion has been characterized by Eqs. (9.1) and (9.2). Thus, in a space admitting an *H-CC* projective motion the Eqs. (8.10), (9.1) and (9.2) are consistent. The following results are concluded by the author et al. [140], Theorems 4.1, 4.2:

Theorem 10.3. The complete integral of an *H-CC* projective motion in an F_n of non-zero curvature is given by

$$p = -(1/4)\,\dot{x}^k\, £\, \mathfrak{B}_k \ln H. \tag{10.4}$$

Proof. Applying the commutation formula (6.7) for the scalar curvature H:

$$(£\, \mathfrak{B}_k - \mathfrak{B}_k\, £)\, H = -(\dot{\partial}_a H)\, £\, G^a_k, \tag{10.5}$$

putting for the Lie derivatives of G^a_k from Eqs. (9.2) and simplifying by means of Eqs. (8.10) and homogeneous property of H, we get the first integral of an *H-CC* projective motion:

$$£\, \mathfrak{B}_k H = -(p\,\dot{\partial}_k H + 2H\, p_k). \tag{10.6}$$

Further, transvecting it by $\dot{x}^k$ and using homogenous properties of the function p, there follows the desired complete integral. //

10.3. Projective curvature collineation

Projective curvature collineations (*P-CC*) in a Finsler space are characterized by Eq. (9.6). In view of Corollary 5.2.1, an isotropic Finsler space F_n $(n > 2)$ has a vanishing projective curvature tensor (to be called *projectively flat*, cf. Chapter 11). Also, a two-dimensional Finsler space F_2 is projectively flat [189]. Thus, the projective curvature tensor is trivially Lie-invariant. Therefore, such spaces are trivial examples of a space with *P-CC*.

As seen in the previous Section (cf. Eqs. (9.8) and (9.9)) the projective tensors W^i_j and W^i_{jk} are also Lie-invariants in the space admitting a *P-CC*. Therefore, we have the:

Theorem 10.4. Either of Eqs. (9.8) and (9.9) represent necessary and sufficient conditions for an infinitesimal motion to define a *P-CC* in an F_n $(n > 2)$.

Further, forming the Lie derivative of the projective deviation tensor given by Eq. (5.2.7):

$$\pounds W^i_j = \pounds H^i_j - \pounds H\, \delta^i_j - \dot{x}^i \pounds (\dot{\partial}_r H^r_j - \dot{\partial}_j H)/(n+1),$$

noting the commutation formula (6.1), and putting from Eqs. (2.9b) and (8.10) holding in a space with *H-CC* above tensor vanishes identically giving rise to a *P-CC*. Thus, we have the:

Theorem 10.5. An *H-CC* is always a *P-CC* in F_n.

In the following we consider a Finsler space of constant Riemannian curvature, say R. The Berwald's tensor H^i_{jk} in this space is given by Eq. (4.11.7). The same, on transvection by $\dot{x}^k$, for Eqs. (4.2.3), (4.9.16) and (4.11.5), yields

$$H^i_j - H\,\delta^i_j = -R\,\dot{x}^i y_j. \tag{10.7}$$

Its directional derivative w.r.t. $\dot{x}^i$, for homogeneous properties of y_j, also determines

$$\dot{\partial}_r H^r_j - \dot{\partial}_j H = -(n+1)\,R\,y_j. \tag{10.8}$$

As such, the projective deviation tensor W^i_j, given by Eq. (5.2.7), vanishes identically for Eqs. (10.7) and (10.8) making the space projectively flat and, therefore, admitting a trivial P-*CC*. Thus, we have proved:

Theorem 10.6. Every infinitesimal transformation defines a trivial *P-CC* in a Finsler space of constant Riemannian curvature.

Considering vanishing Lie derivative of the projective entities Q^i_{jkh} introduced by the present author [98], Pandey [172] defines projective curvature collineations, and called these as "*special projective curvature collineations*".

§ 11. Conformal and homothetic motions in F_n with reference to Lie Derivation

Conformal and homothetic motions were dealt in § 7 of Chapter 5. Presently, we revisit that discussion in relation with the concept of Lie derivation w.r.t. an infinitesimal transformation η given by Eq. (5.4.1). Writing the conformal change h_{ij}, given by Eq. (5.7.4), as $\pounds\, g_{ij}$ the necessary conditions for a conformal motion are [234], Eqs. (3.1.11):

$$\pounds\, g_{ij} = \psi\, g_{ij}, \tag{11.1}$$

where the scalar function ψ is a point function given by Eq. (5.3.4).

CHAPTER 7

SYMMETRIC AND PROJECTIVELY SYMMETRIC FINSLER SPACES

§ 1. Introduction

The concept of a symmetric space in Riemannian geometry was introduced by E. Cartan [31] in terms of vanishing covariant derivative of its Riemannian curvature tensor. This theory was extended to Finslerian geometry by the author [108] taking vanishing covariant derivative of the non-null Berwald's curvature tensor:

$$\mathfrak{B}_l \, H^i_{ikh} = 0. \tag{1.1}$$

An n-dimensional Finsler space equipped with above condition is called a *symmetric Finsler space* and is denoted by $S\text{-}F_n$. Some examples of such spaces and necessary consequences of the condition expressed by Eq. (1.1) are considered. Reduction of some identities of geometrical importance such as commutation formulae given by Eqs. (4.7.14), (4.8.3), etc. and Lie-invariance of Berwald's scalar curvature are dealt with. A symmetric Finsler space with a concircular vector field generating the infinitesimal transformation is considered by Pandey [179].

This concept of symmetry was further extended to Riemannian spaces by Gy. Soós [217] and to Finsler spaces by the present author taking vanishing covariant derivative of the projective curvature tensors of the respective spaces. Such spaces are called *projectively symmetric* Finsler spaces and are denoted by $PS\text{-}F_n$ [109]. Examples of such spaces, conditions for an $S\text{-}F_n$ to become a $PS\text{-}F_n$, reduction of various identities, and the decomposed $PS\text{-}F_n$ are considered in §§ 5 - 8.

Later, the author considered that class of Finsler spaces where the second order covariant derivative of the Berwald's curvature tensor vanishes [139]. Such Finsler spaces are called *bi-symmetric* and are denoted by $HS^2\text{-}F_n$.

1.1. Affine motion in $S\text{-}F_n$ and $PS\text{-}F_n$

Affine motions generated by vector fields of different kinds discussed in § 9 of Chapter 5 are studied by the author et al. for $S\text{-}F_n$ in [96]; while such motions generated by a recurrent vector field (cf. Eq.

(5.9.1d)) for S-F_n are studied by the author et al. in [108]. Affine motions generated by contra-vector field (cf. Eq. (5.9.1a) are studied in PS-F_n by the author with Meher [134].

1.2. Projective motions in S-F_n and PS-F_n

Meher [88] studied projective motions in S-F_n while the author with Pandey [141] studied projective motions in a Finsler space characterized by vanishing covariant derivative of the normal projective curvature tensor N^{i}_{jkh} (cf. Eq. (11.6.1) below). Pande and Kumar discussed some special cases of projective motion in S-F_n [166]. Pandey [180] also supplemented the study of projective motions for both S-F_n and PS-F_n.

§ 2. Some immediate implications of Eq. (1.1)

Successive transvections of Eq. (1.1) by $\dot{x}^i$'s, for Eqs. (4.6.7c), (4.9.15b) and (4.9.16) also yield

$$\mathfrak{B}_l\, H^{i}_{jk} = 0. \qquad (2.1); \qquad \mathfrak{B}_l\, H^{i}_{j} = 0. \qquad (2.2)$$

Also, contractions of these equations w.r.t. the indices i and j, for Eqs. (4.9.20), give

$$\text{(a)}\ \mathfrak{B}_l\, H_{kh} \equiv 0, \quad \text{(b)}\ \mathfrak{B}_l\, H_k \equiv 0, \quad \text{(c)}\ \mathfrak{B}_l\, H = 0. \qquad (2.3)$$

Thus, we have the

Theorem 2.1. Tensors associated to Berwald's curvature tensor, vector H_k and the scalar curvature H are all covariant constants in a S-F_n.

In the following, we derive a necessary and sufficient condition for an F_n satisfying condition (2.1) to become a S-F_n. Applying the commutation formula (4.8.3) to the curvature tensor H^{i}_{jk}:

$$(\dot{\partial}_h\, \mathfrak{B}_l - \mathfrak{B}_l\, \dot{\partial}_h) H^{i}_{jk} = H^{m}_{jk}\, G^{i}_{hlm} + 2H^{i}_{m[j}\, G^{m}_{k]hl}, \qquad (2.4)$$

and putting from Eqs. (4.9.15a) and (2.1), there results

$$-\mathfrak{B}_l\, H^{i}_{jkh} = H^{m}_{jk}\, G^{i}_{hlm} + 2H^{i}_{m[j}\, G^{m}_{k]hl}. \qquad (2.5)$$

Thus, the space admitting Eq. (2.1) becomes symmetric if the expression on the RHS of above equation vanishes identically:

$$H^m_{jk}\, G^i_{hlm} + 2H^i_{m\,[j}\, G^m_{k]hl} = 0. \tag{2.6}$$

For a symmetric space having Eqs. (1.1) and, therefore, (2.1) also, the Eq. (2.6) follows as a necessary condition from Eq. (2.5). On contrary, when Eq. (2.6) holds in the space admitting Eq. (2.1), the sufficiency of the condition to turn the space symmetric follows from Eq. (2.5). Hence, we have the:

Theorem 2.2. A Finsler space F_n with a covariant constant H^i_{jk} becomes symmetric F_n if and only if there holds Eq. (2.6).

Contraction of Eq. (2.6) w.r.t. the indices i and j, for Eqs. (4.9.20), also yields

$$H_m\, G^m_{khl} = 0, \tag{2.7}$$

in a symmetric space.

The Berwald's tensor H^i_{jk} in an isotropic Finsler space F_n, ($n > 2$), of constant Riemannian curvature R is given by Eq. (4.11.7). Its covariant derivative, for Eq. (4.6.10), vanishes identically. Hence, we have the:

Lemma 2.1. In an isotropic Finsler space F_n, ($n > 2$), of constant Riemannian curvature R the Berwald's tensor H^i_{jk} is a covariant constant.

In view of this lemma and Theorem 2.2, there also follows:

Theorem 2.3. An isotropic Finsler space F_n, ($n > 2$), of constant Riemannian curvature R admits the condition in Eq. (2.6) if and only if it is a symmetric space.

Affinely-connected Finsler space is characterized by Eq. (4.6.5) leading to independence of Berwald's connection parameters on the directional arguments [189], p. 81:

$$G^i_{hlm} = 0. \tag{2.8}$$

Thus, the condition in Eq. (2.6) is obviously satisfied in such a space and Theorem 2.3 gives rise to the:

Theorem 2.4. An isotropic Finsler space F_n, ($n > 2$), of constant Riemannian curvature is necessarily a symmetric space.

§ 3. Some more identities in an S-F_n

For the vanishing characteristics of the covariant derivative of the curvature tensor and its associated fields several identities can be simplified in an S-F_n . For instance, the (generalized) Bianchi's second identity given by Eq. (4.10.5), for Eq. (1.1), reduces to

$$G^i_{lm[j} H^l_{kh]} = 0. \tag{3.1}$$

For skew-symmetric properties of the tensor H^l_{kh} in its lower indices, above identity expands to

$$G^i_{lmj} H^l_{kh} + G^i_{lmk} H^l_{hj} + G^i_{lmh} H^l_{jk} = 0.$$

Its transvection with $\dot{x}^k$, for Eqs. (4.9.16) and (4.10.15), yields

$$-G^i_{lmj} H^l_h + G^i_{lmh} H^l_j \equiv 2G^i_{lm[h} H^l_{j]} = 0. \tag{3.2}$$

Next, the commutation formula given by Eq. (4.7.14), when applied to the Berwald's curvature tensor H^i_{hlm} and its associated tensors H^i_{hl}, H^i_h, for Eqs. (1.1), (2.1) and (2.2), yields

$$H^i_{jkp} H^p_{hlm} + 2H^p_{jk[h} H^i_{l]pm} = H^p_{jkm} H^i_{hlp} + (\dot{\partial}_p H^i_{hlm}) H^p_{jk}, \tag{3.3}$$

$$H^i_{jkp} H^p_{hl} + 2H^p_{jk[h} H^i_{l]p} = H^p_{jk} H^i_{hlp}, \tag{3.4}$$

$$H^i_{jkp} H^p_h - H^p_{jkh} H^i_p = H^p_{jk} (\dot{\partial}_p H^i_h). \tag{3.5}$$

Finally, formulae for Lie derivatives of the curvature tensor and its associated tensors, obtainable from Eq. (5.2.16b), can be also reduced in the light of Eqs. (1.1), (2.1) and (2.2) for an S-F_n. Also, the Lie derivative of the Berwald's scalar curvature H, found by Eq. (5.2.3b):

$$\pounds H = (\partial_j H) v^j + (\dot{\partial}_j H)(\partial_k v^j)\dot{x}^k = v^j \mathfrak{B}_j H + (\dot{\partial}_k H)(\mathfrak{B}_h v^k)\dot{x}^h, \tag{3.6}$$

for Eq. (2.3c), reduces to

$$\pounds H = (\dot{\partial}_k H)(\mathfrak{B}_h v^k)\dot{x}^h, \tag{3.7}$$

in an S-F_n . This proves the:

Theorem 3.1. In an $S\text{-}F_n$ the scalar curvature H is Lie-invariant if and only if it is independent of the directional arguments.

§ 4. Affine motion in an $S\text{-}F_n$

The Eq. (6.8.16), for an $S\text{-}F_n$ having Eq. (2.1), reduces to

$$(\mathfrak{B}_h v^a) H^i_{a\,j} + \mathfrak{B}_h (\mathfrak{B}_j \mathfrak{B}_k v^i \dot{x}^k) = 0. \tag{4.1}$$

For affine motion generated by a recurrent vector field satisfying Eq. (5.9.1d) and, therefore, Eq. (6.8.16c), above equation further reduces to:

$$\mu_h v^a H^i_{a\,j} + \mathfrak{B}_h (\mu_{jk} v^i \dot{x}^k) = 0. \tag{4.2}$$

Its contraction w.r.t. indices i and j, for Eq. (6.8.14b) holding in F_n and, therefore, also in $S\text{-}F_n$, makes the second term zero leaving the equation:

$$\mu_h v^a H^i_{a\,i} = -\mu_h v^a H_a = 0, \tag{4.3}$$

by Eq. (4.9.20). The author concluded Theorem 4.1 in [108] from above equation; but, it may be noted that the condition expressed by Eq. (4.5 ii) in [108] is already satisfied for an affine motion generated by a recurrent vector field in a general Finsler space F_n (cf. Eq. (6.8.15c)). Hence, the vector μ_h in above equation need not be zero which is also a requirement for the recurrent affine motion (cf. Eq. (5.9.1d)).

§ 5. Projectively symmetric Finsler spaces

Definition 5.1. A Finsler space having vanishing covariant derivative of its projective curvature tensor W^i_{jkh} is called a projectively symmetric Finsler space and is denoted by $PS\text{-}F_n$:

$$\mathfrak{B}_l W^i_{jkh} = 0. \tag{5.1}$$

As in § 2, the successive transvections of Eq. (5.1) by $\dot{x}^h$, for Eqs. (4.6.7), (5.2.16) and (5.2.17) also yield

$$\mathfrak{B}_l W^i_{jk} = 0 \quad (5.2); \qquad \text{and} \qquad \mathfrak{B}_l W^i_j = 0. \quad (5.3)$$

In view of Corollary 5.2.1 and Note 5.2.1, Eqs. (5.1) - (5.3) are evidently satisfied in an F_2. Hence, we have the

Example 5.1. An isotropic Finsler space is projectively symmetric.

Example 5.2. The F_2 is always projectively symmetric.

It may be noted from Eqs. (5.2) and (5.3) that the tensors W^i_{jk} and W^i_j are necessarily covariant constants in a PS-F_n. However, the converse need not be true. In the following, we derive a sufficient condition to transform a F_n equipped with Eq. (5.2) into a PS-F_n.

The directional derivative of Eq. (5.2), for the commutation formula (4.8.3), yields

$$\mathfrak{B}_l \, \dot{\partial}_h W^i_{jk} + W^m_{jk} G^i_{mlh} - W^i_{mk} G^m_{jlh} - W^i_{jm} G^m_{klh} = 0,$$

or, for Eqs. (5.2.14)

$$\mathfrak{B}_l \, W^i_{jkh} + W^m_{jk} G^i_{mlh} + 2W^i_{m[j} G^m_{k]lh} = 0, \tag{5.4}$$

which reduces to Eq. (5.1) if there holds the condition

$$W^m_{jk} G^i_{mlh} + 2W^i_{m[j} G^m_{k]lh} = 0. \tag{5.5}$$

Thus, we have the

Theorem 5.1. A Finsler space F_n ($n > 2$) admitting Eq. (5.2) becomes projectively symmetric if and only if there holds the condition vide Eq. (5.5).

It is seen in § 6, Chapter 4 that the connection parameters are independent of the directional arguments leading to zero tensor G^i_{jkh}. As such, the condition in Eq. (5.5) holds in such a space. Thus, we have

Corollary 5.1. An affinely connected Finsler space admitting Eq. (5.2) is a PS-F_n.

In the following, we examine if the scalar curvature H is constant in a PS-F_n? Plugging in the value of projective deviation tensor given by Eq. (5.2.7) in Eq. (5.3), we get

$$\mathfrak{B}_l \, H^i_j - \mathfrak{B}_l \, H \, \delta^i_j = \dot{x}^i \, \mathfrak{B}_l \, (\dot{\partial}_r H^r_j - \dot{\partial}_j H) / (n+1). \tag{5.6}$$

On contraction w.r.t. i and j, for Eq. (4.6.7), (4.8.3), (4.9.20) and homogeneous properties of *H*, it yields the identity

$$-\mathfrak{B}_l H = \{\dot{x}^i \mathfrak{B}_l \dot{\partial}_r H^r_i - 2\mathfrak{B}_l H\} / (n+1)\},$$

or,
$$(1-n)\mathfrak{B}_l H = \dot{x}^i \mathfrak{B}_l \dot{\partial}_r H^r_i. \tag{5.8}$$

Thus, in general, *H* is not constant in *PS*-F_n unless the tensor H^r_i; and, therefore, the scalar *H* itself are independent of the directional arguments.

Transvection of the identity (5.5) by $\dot{x}^k$, for Eqs. (4.10.15) and (5.2.17), also gives

$$W^m_j G^i_{mlh} - W^i_m G^m_{jlh} = 0. \tag{5.9}$$

5.1. Applying the commutation rule given by Eq. (4.7.14) to the projective tensors W^i_{jkh}, W^i_{jk} and W^i_j in *PS*-F_n, there follow the identities

$$H^i_{lmp} W^p_{jkh} + 2H^p_{lm[j} W^i_{k]ph} = H^p_{lmh} W^i_{jkp} + H^p_{lm} \dot{\partial}_p W^i_{jkh}, \tag{5.9}$$

$$H^i_{lmp} W^p_{jk} + 2H^p_{lm[j} W^i_{k]p} = H^p_{lm} \dot{\partial}_p W^i_{jk}, \tag{5.10}$$

$$H^i_{lmp} W^p_j - H^p_{lmj} W^i_p = H^p_{lm} \dot{\partial}_p W^i_j. \tag{5.11}$$

Note 5.1. The last two identities are also directly derivable from Eq. (5.9) on its successive transvections with $\dot{x}^h$, $\dot{x}^k$ and making use of Eqs. (4.9.15b), (5.2.16), (5.2.17) and homogeneous properties of W^i_{jkh}.

§ 6. Possibilities for a *S*-F_n to become projectively symmetric

A symmetric Riemannian space is seen to be projectively symmetric [72]. On contrary, the author [109] has proved that it is not the case with a *S*-F_n to be *PS*-F_n in general.

For the Eqs. (1.1) and (2.1) - (2.3) holding in a *S*-F_n, it can be easily verified from Eq. (5.2.13c) that the covariant derivative of the projective tensor W^i_{jk} vanishes identically and satisfies Eq. (5.2). Further, a transvection of Eq. (5.2) by $\dot{x}^k$, for Eqs. (4.6.7) and (5.2.17), also yields Eq. (5.3) in the space.

Thus, there holds the:

Theorem 6.1. The projective tensors W^i_j and W^i_{jk} are covariant constants in a symmetric Finsler space.

As seen in Theorem 5.1, the Eq. (5.5) represents a necessary and sufficient condition for a F_n admitting Eq. (5.2) to become a *PS*-F_n. Thus, the Theorem 5.1 gives rise to the:

Theorem 6.2. A *S*-F_n becomes a *PS*-F_n if and only if it admits Eq. (5.5).

6.1. Alternate form of Eq. (5.5)

Putting for the projective tensor W^i_{jk} given by Eq. (5.2.13c) in Eq. (5.5), and simplifying by means of Eq. (4.10.15), there follows

$$H^p_{jk}\, G^i_{plm} + 2\dot{x}^i\, G^p_{lm\,[k}\, H^r_{j]\,pr} /(n+1)$$

$$+2\{H^i_{p\,[j} + (\dot{x}^r\, H_{rp} + nH_p)\,\delta^i_{[j} /(n^2-1)\}\, G^p_{k]\,lm} = 0. \qquad (6.1)$$

For Eqs. (2.6) and (2.7) holding in a *S*-F_n, above equation reduces to

$$\dot{x}^i\, G^p_{lm\,[k}\, H^r_{j]\,pr} + \{\dot{x}^r\, H_{rp} / (n-1)\}\, \delta^i_{[j}\, G^p_{k]\,lm} = 0,$$

or,
$$G^p_{lm\,[j}\, \{\dot{x}^i\, H^r_{k]\,pr} + \delta^i_{k]}(\dot{x}^r\, H_{rp}) / (n-1)\} = 0. \qquad (6.2)$$

On the other hand, the covariant derivative of Eq. (5.2.15), for Eqs. (4.6.7), (1.1) and (2.3), simplifies to

$$\mathfrak{B}_l\, W^i_{jkh} = -\, \dot{x}^i\, \mathfrak{B}_l\, (\dot{\partial}_h\, H^r_{jkr}) /\, (n+1)$$

$$+\, 2\, (n\, \mathfrak{B}_l\, \dot{\partial}_h\, H_{[j} + \dot{x}^r\, \mathfrak{B}_l\, \dot{\partial}_h\, H_{r\,[j})\, \delta^i_{k]} / (n^2 - 1).$$

Application of the commutation formula (4.8.3) for above Berwald's tensors and use of Eqs. (4.10.15), (1.1), (2.3) and (2.7) reduces above equation to

$$\mathfrak{B}_l\, W^i_{jkh} = \frac{2}{n+1} G^p_{lh\,[j} \left\{\dot{x}^i\, H^r_{k]\,pr} + \frac{1}{n-1} \delta^i_{k]} (\dot{x}^r\, H_{rp})\right\}, \qquad (6.3)$$

which vanishes for Eq. (6.2) making *S*-F_n projectively symmetric. This establishes the

Theorem 6.3. A symmetric Finsler space is projectively symmetric if and only there holds the Eq. (6.2).

Further, an affinely connected Finsler space having vanishing tensor G^i_{jkh} satisfies the condition (6.2). Thus, analogous to Corollary 5.1, there also holds the:

Corollary 6.1. An affinely connected symmetric Finsler space is also projectively symmetric.

§ 7. Bianchi identities in a *PS-F$_n$*

Forming the covariant derivative of Eq. (5.2.15) w.r.t. x^l and interchanging the processes of derivation as per commutation formula (4.8.3) applied to the tensors therein, the skew-symmetric part of the resulting equation w.r.t. the indices l, j, k yields the generalization of Bianchi's second identity

$$\mathfrak{B}_{[l} W^i_{jk]h} = \mathfrak{B}_{[l} H^i_{jk]h} - \{\mathfrak{B}_{[l} H^r_{jk]\,r}\,\delta^i_h + \dot{x}^i \dot{\partial}_h\, \mathfrak{B}_{[l} H^r_{jk]\,r}\}/(n+1)$$

$$+\,2(n\,\dot{\partial}_h \mathfrak{B}_{[l}\, H_j + \mathfrak{B}_{[l}\, H_{|h|j} + \dot{x}^r \dot{\partial}_h\ {}_{[l}\, H_{|r|j})\,\delta^i_{k]}\,/\,(n^2-1), \quad (7.1)$$

In the following we check this identity in the symmetric and projectively symmetric Finsler spaces. First, in *S-F$_n$* having Eqs. (1.1), (2.3), it reduces to

$$\mathfrak{B}_{[l} W^i_{jk]h} = 0. \quad (7.2)$$

Its transvection with the vector $\dot{x}^h$, for Eqs. (4.6.7) and (5.2.16), also determines

$$\mathfrak{B}_{[l} W^i_{jk]} = 0, \quad (7.3)$$

which, by Theo. 6.1, gets term wise satisfied in *S-F$_n$*. Further, re-writing the terms in Eq. (7.2) as

$$\mathfrak{B}_{[l}\, \dot{\partial}_{|h|}\, W^i_{jk]} = 0,$$

and interchanging the operators of derivation according to the commutation formula (4.8.3), the Eq. (7.2) yields

$$\dot{\partial}_h \mathfrak{B}_{[l} W^i_{jk]} - W^m_{[jk} G^i_{l]\,mh} = 0;$$

which, for Eq. (7.3) further reduces to

$$W^{m}_{[jk}\, G^{i}_{l]\,mh} = 0. \tag{7.4}$$

Putting from Eq. (5.2.13c), and simplifying by means of Eq. (4.10.15), it transforms to

$$H^{m}_{[jk} G^{i}_{l]\,mh} = 0; \tag{7.5}$$

which is also evident from the Bianchi identity given by Eq. (4.10.5) in a $S\text{-}F_n$. Thus, there hold the identities given by Eqs. (7.2) - (7.5) in a $S\text{-}F_n$.

It is to be noted that, for Eqs. (5.1) and (5.2), the identities given by Eqs. (7.2) and (7.3) also hold in a $PS\text{-}F_n$. In addition, the identity vide Eq. (5.5) is already seen to hold in $PS\text{-}F_n$. Forming its skew-symmetric part w.r.t. the indices l, j, k, there also results the identity (7.4) and, therefore, Eq. (7.5) as seen above. Thus, the identities in Eqs. (7.2) - (7.5) also hold good in a $PS\text{-}F_n$. These remarks give rise to the:

Theorem 7.1. The identities vide Eqs. (7.2) - (7.5) hold good in either of the symmetric and projective symmetric Finsler spaces.

The identity vide Eq. (7.5) - holding in a $PS\text{-}F_n$ reduces the Bianchi identity given by Eq. (4.10.5)

$$\mathfrak{B}_{[l} H^{i}_{\ jk]h} = 0. \tag{7.6}$$

Consequently, the identity (7.1) reduces to

$$(n\dot{\partial}_h \mathfrak{B}_{[l}\, H_j + \mathfrak{B}_{[l}\, H_{|h|j} + \dot{x}^r \dot{\partial}_h\, \mathfrak{B}_{[l}\, H_{|r|j})\, \delta^i_{k]} = 0. \tag{7.7}$$

Its transvection with the vector $\dot{x}^h$, for homogenous properties of the Berwald's tensors also yields

$$(n\,\mathfrak{B}_{[l}\, H_j + \dot{x}^h \mathfrak{B}_{[l}\, H_{|h|j})\, \delta^i_{k]} = 0. \tag{7.8}$$

Thus, we conclude the:

Theorem 7.2. There also hold the identities vide Eqs. (7.6) - (7.8) in a $PS\text{-}F_n$.

§ 8. Some more identities in *PS*-F_n involving Lie derivation

Applying the formula (6.2.16b) to the projective tensors W^{i}_{jkh}, W^{i}_{jk} and W^{i}_{j} in $PS\text{-}F_n$, there follow the identities

$$\pounds W^i_{jkh} = W^i_{jkl}\, \mathfrak{B}_h v^l - W^l_{jkh} \mathfrak{B}_l\, v^i - 2(\mathfrak{B}_{[j}\, v^l)\, W^i_{k]lh}$$

$$+ (\dot{\partial}_l\, W^i_{jkh})\, (\mathfrak{B}_m v^l)\, \dot{x}^m, \tag{8.1}$$

$$\pounds W^i_{jk} = W^i_{jkl}\, (\mathfrak{B}_h v^l)\, \dot{x}^h - W^l_{jk} \mathfrak{B}_l\, v^i - 2(\mathfrak{B}_{[j}\, v^l)\, W^i_{k]l}, \tag{8.2}$$

$$\pounds W^i_j = (\dot{\partial}_l\, W^i_j)(\mathfrak{B}_h v^l)\, \dot{x}^h - W^l_j \mathfrak{B}_l\, v^i + W^i_l\, \mathfrak{B}_j\, v^l. \tag{8.3}$$

Note 8.1. Like Note 5.1, the last two identities are directly derivable on successive transvection of Eq. (8.1) by $\dot{x}^h$ and $\dot{x}^k$, for Eqs. (6.2.9b), (5.2.16), (5.2.17) and the homogeneous properties of the tensor W^i_{jkh}.

It is seen in the previous chapter (cf. Eqs. (6.9.6), (6.9.8) and (6.9.9)) that above projective tensors are Lie invariants in a space admitting a projective motion. Thus, for a $PS\text{-}F_n$ admitting a projective motion, the necessary conditions follow from Eqs. (8.1) - (8.3):

$$W^i_{jkl} \mathfrak{B}_h v^l + (\dot{\partial}_l\, W^i_{jkh})\, \mathfrak{B}_m v^l\, \dot{x}^m = W^l_{jkh} \mathfrak{B}_l v^i + 2(_{[j}\, v^l)\, W^i_{k]lh}, \tag{8.4}$$

$$W^i_{jkl}\, (\mathfrak{B}_h v^l)\, \dot{x}^h = W^l_{jk} \mathfrak{B}_l\, v^i + 2(\mathfrak{B}_{[j}\, v^l)\, W^i_{k]l}, \tag{8.5}$$

$$(\dot{\partial}_l\, W^i_j)(\mathfrak{B}_h v^l)\, \dot{x}^h = W^l_j \mathfrak{B}_l\, v^i - W^i_l\, \mathfrak{B}_j\, v^l. \tag{8.6}$$

§ 9. A decomposable *PS-F*$_n$

We consider a Finsler space F_n with a decomposed metric

$$(ds)^2 = g_{\alpha\beta}(x^\varepsilon, \dot{x}^\varepsilon)\, dx^\alpha\, dx^\beta + g_{\lambda\mu}(x^\sigma, \dot{x}^\sigma)\, dx^\lambda\, dx^\mu, \tag{9.1}$$

where the early Greek indices α, β, γ, ... run over positive integral values from 1 to r while the latter ones λ, μ, ν, ... assume values from $r + 1$ to n and the decomposed components of the metric tensor: $g_{\alpha\beta}$ and $g_{\lambda\mu}$ are the functions of variables $(x^\varepsilon, \dot{x}^\varepsilon)$ and $(x^\sigma, \dot{x}^\sigma)$ only. It may be noticed that the connection parameters G^i_{jk} are decomposed into $G^\alpha_{\beta\gamma}(x^\varepsilon, \dot{x}^\varepsilon)$ and $G^\lambda_{\mu\nu}(x^\sigma, \dot{x}^\sigma)$ containing variables of the decomposed ranges only. Remaining components of G^i_{jk} vanish identically. Similar is the case with the components of the curvature tensor H^i_{jkh}

and its associate tensors. But, the projective curvature tensor decomposes according to

$$\left.\begin{aligned}
W^{i}_{jkh}(x^{l},\dot{x}^{l}) &= W^{\alpha}_{\beta\gamma\delta}(x^{\varepsilon},\dot{x}^{\varepsilon})\ , \text{ for } 1 \le i,j,k,h \le r;\\
W^{\alpha}_{\mu\gamma\rho}(x^{\sigma},\dot{x}^{\sigma}) &= (nH_{\mu\rho} + H_{\rho\mu} + \dot{x}^{\sigma}\dot{\partial}_{\rho}H_{\sigma\mu})\,\delta^{\alpha}_{\gamma}/(n^{2}-1);\\
W^{\alpha}_{\mu\nu\delta}(x^{\sigma},\dot{x}^{\sigma}) &= H^{\sigma}_{\nu\mu\sigma}\,\delta^{\alpha}_{\delta}/(n+1);\\
W^{\alpha}_{\mu\nu\rho}(x^{i},\dot{x}^{i}) &= \dot{x}^{\alpha}\dot{\partial}_{\rho}H^{\sigma}_{\nu\mu\sigma}/(n+1);\\
W^{\lambda}_{\beta\gamma\delta}(x^{i},\dot{x}^{i}) &= \dot{x}^{\lambda}\dot{\partial}_{\delta}H^{\varepsilon}_{\gamma\beta\varepsilon}/(n+1);\\
W^{\lambda}_{\beta\gamma\rho}(x^{\sigma},\dot{x}^{\sigma}) &= H^{\varepsilon}_{\gamma\beta\varepsilon}\,\delta^{\lambda}_{\rho}/(n+1);\\
W^{\lambda}_{\beta\nu\delta}(x^{\sigma},\dot{x}^{\sigma}) &= (n\,H_{\beta\delta} + H_{\delta\beta} + \dot{x}^{\varepsilon}\dot{\partial}_{\delta}H_{\varepsilon\beta})\,\delta^{\lambda}_{\nu}/(n^{2}-1);\\
W^{i}_{jkh}(x^{l},\dot{x}^{l}) &= W^{\lambda}_{\mu\nu\rho}(x^{\sigma},\dot{x}^{\sigma})\ , \text{ for } r < i,j,k,h \le n.
\end{aligned}\right\} \quad (9.2)$$

Rest components of W^{i}_{jkh} are zero or linearly dependent on above.

Also, the non-zero and linearly independent components of the covariant derivative of W^{i}_{jkh} are given by

$$\left.\begin{aligned}
\mathfrak{B}_{l}\,W^{i}_{jkh}(x^{l},\dot{x}^{l}) &= \mathfrak{B}_{\varepsilon}W^{\alpha}_{\beta\gamma\delta}(x^{\varepsilon},\dot{x}^{\varepsilon}), \text{ for } 1 \le i,j,k,h \le r;\\
\mathfrak{B}_{\lambda}W^{\alpha}_{\mu\gamma\rho}(x^{\sigma},\dot{x}^{\sigma}) &= \mathfrak{B}_{\lambda}(nH_{\mu\rho} + H_{\rho\mu} + \dot{x}^{\sigma}\dot{\partial}_{\rho}H_{\sigma\mu})\,\delta^{\alpha}_{\gamma}/(n^{2}-1);\\
\mathfrak{B}_{\lambda}\,W^{\alpha}_{\mu\nu\delta}(x^{\sigma},\dot{x}^{\sigma}) &= \mathfrak{B}_{\lambda}\,H^{\sigma}_{\nu\mu\sigma}\,\delta^{\alpha}_{\delta}/(n+1);\\
\mathfrak{B}_{\lambda}\,W^{\alpha}_{\mu\nu\rho}(x^{i},\dot{x}^{i}) &= \dot{x}^{\alpha}\,\mathfrak{B}_{\lambda}\dot{\partial}_{\rho}H^{\sigma}_{\nu\mu\sigma}/(n+1);\\
\mathfrak{B}_{\alpha}W^{\lambda}_{\beta\gamma\delta}(x^{i},\dot{x}^{i}) &= \dot{x}^{\lambda}\,\mathfrak{B}_{\alpha}\dot{\partial}_{\delta}H^{\varepsilon}_{\gamma\beta\varepsilon}/(n+1);\\
\mathfrak{B}_{\alpha}W^{\lambda}_{\beta\gamma\rho}(x^{\sigma},\dot{x}^{\sigma}) &= \mathfrak{B}_{\alpha}H^{\varepsilon}_{\gamma\beta\varepsilon}\,\delta^{\lambda}_{\rho}/(n+1);\\
\mathfrak{B}_{\alpha}W^{\lambda}_{\beta\nu\delta}(x^{\sigma},\dot{x}^{\sigma}) &= \mathfrak{B}_{\alpha}(n\,H_{\beta\delta} + H_{\delta\beta} + \dot{x}^{\varepsilon}\dot{\partial}_{\delta}H_{\varepsilon\beta})\,\delta^{\lambda}_{\nu}/(n^{2}-1);\\
\mathfrak{B}_{l}\,W^{i}_{jkh}(x^{l},\dot{x}^{l}) &= \mathfrak{B}_{\sigma}W^{\lambda}_{\mu\nu\rho}(x^{\sigma},\dot{x}^{\sigma}), \text{ for } r < i,j,k,h \le n.
\end{aligned}\right\} \quad (9.3)$$

If the decomposed space be projectively symmetric above covariant derivatives vanish yielding the following results. The third and fourth of Eqs. (9.3) give

$$\mathfrak{B}_\lambda H^{\sigma}_{\nu\,\mu\sigma} = 2\mathfrak{B}_\lambda H_{[\mu\nu]} = 0$$

$\Rightarrow$

$$\mathfrak{B}_\lambda H_{\mu\nu} = {}_\lambda H_{\nu\mu}, \tag{9.4a}$$

and

$$\mathfrak{B}_\lambda \dot{\partial}_\rho H^{\sigma}_{\nu\,\mu\sigma} = 0, \text{ for arbitrary } \dot{x}^\alpha, \tag{9.4b}$$

respectively. Its transvection by $\dot{x}^\nu$, for Eqs. (4.6.7), (4.9.21b), (4.10.3), and homogeneous properties of the tensor $H_{\sigma\mu}$, also yields

$$\mathfrak{B}_\lambda \dot{x}^\nu \dot{\partial}_\rho H_{\nu\mu} = 0. \tag{9.5}$$

Hence, Eq. (9.3ii), for Eqs. (9.4a) and (9.5), yields

$$\mathfrak{B}_\lambda H_{\mu\rho} = 0. \tag{9.6}$$

Similarly, from the fifth, sixth and seventh of Eqs. (9.3), we may derive

$$\mathfrak{B}_\alpha H_{\beta\delta} = 0. \tag{9.7}$$

For Eq. (4.10.3), above Eqs. (9.6) and (9.7) also yield

$$\mathfrak{B}_\lambda H^{\sigma}_{\mu\rho\,\sigma} = 0 \quad (9.8); \quad \text{and} \quad \mathfrak{B}_\alpha H^{\varepsilon}_{\beta\delta\varepsilon} = 0. \tag{9.9}$$

Hence, we conclude the:

Theorem 9.1. The contracted curvature tensors of Berwald are covariant constants in a *PS*-F_n with decomposed metric vide Eq. (9.1).

§ 10. Affine motion in a *PS*-F_n

Affine motions in a space F_n are characterized by necessary and sufficient conditions vide Eq. (6.8.1) with their integrability conditions given by Eqs. (6.8.5) - (6.8.7), (6.8.10) and (6.8.12). Affine motions generated by a special vector fields in F_n were also discussed in Chapter 6, Sub-section 8.2. The study of the affine motions generated by special concircular [30)] vector field given by Eq. (5.9.1c) was pursued by the author with Meher [134]. Such affine motions were briefly denoted by *CA*-motions.

The Lie derivative of the projective tensor W^i_{jk}, in $PS\text{-}F_n$ is obtained vide Eq. (8.2) and it vanishes for an affine motion in view of Eq. (6.8.12). The same equation (8.2), for a special concircular affine motion characterized by Eq. (5.9.1c), simplifies to

$$W^i_{jk} = 0. \tag{10.1}$$

This concludes the:

Theorem 10.1. A projectively symmetric Finsler space admitting a special concircular affine motion is necessarily projectively flat.

Foot-note:

[30)] However, it was called a contra vector field and the affine motion generated by it as contra affine (or *CA*-) motion in [134].

CHAPTER 8

RECURRENT FINSLER SPACES

§ 1. Introduction

As mentioned in Chapter 2, the study of Riemannian spaces of recurrent curvature, introduced by A.G. Walker [230] has been extended to Finsler geometry and to the generalized metric and non-metric spaces. Finslerian spaces of recurrent curvature have been introduced by Moór [147] and are studied by many others [65], [97], [196], [210], [211], [213] including the present author [110], and the author with Meher [130], [133]. An n-dimensional Finsler space with Berwald's curvature tensor satisfying the recurrence property:

$$\mathfrak{B}_l H^i_{jkh} = \lambda_l H^i_{jkh}, \tag{1.1}$$

for some non-null vector field λ_l has been called a "*recurrent Finsler space*" and is denoted by HR- F_n by the present author and Meher [130]. Sinha [213], [214] and the present author with Meher [130] studied the existence of affine motions in recurrent Finslerian spaces. An HR-F_n admitting an affine motion has been denoted by AHR-F_n [130]. As a necessary consequence of the existence of affine motion in such a space certain properties of the curvature tensor and that of a tensor

$$A_{lm} \equiv 2\,\mathfrak{B}_{[l}\,\lambda_{m]}, \tag{1.2}$$

have been derived by the present author and Meher [op. cit.].

Also, the present author and Meher [130] studied recurrent affine motions (i.e. equipped with the condition (5.9.1d) on their generating vector field v^i) in recurrent Finsler space HR-F_n. An HR-F_n admitting an infinitesimal transformation given by Eq. (5.4.1) of above type is called a "special HR-F_n" and is denoted by SHR-F_n [128]. Certain necessary consequences of the existence of such recurrent affine motions and their integrability conditions have been derived therein

Kumar [65] made an analogous study of affine motions in recurrent Finslerian spaces employing Cartan's curvature tensors. Imposing an additional condition of vanishing Lie derivative of the recurrence vector, Pandey [176] studied recurrent affine motions in an HR-F_n.

All above authors took the recurrence vector as a function of line-elements. Remarkably enough, it was noticed independently by Meher [89] and Pandey [177] that the recurrence vector λ^i in *HR*-F_n of non-zero sectional curvature H is necessarily independent of directional arguments. A direct and shorter proof of this observation was also given by the present author [112]. Independence of λ^i upon the directional coordinates makes a serious deviation in the existing theory of recurrent Finsler spaces. In the light of this observation, study of affine motions has been revised by the present author [112] and Pandey [177]. The author further revised the theory of projective motions carried out earlier [127], [131], [132] in recurrent Finsler spaces in his next work [113].

The study of recurrent Finsler spaces is very vast and needs an independent treatise to cover up. However, the author confines here to present some of the results in such spaces mainly included in author's own works [112], [113].

§ 2. *HR*-F_n with non-zero scalar curvature

Successive transvections of Eq. (1.1) by $\dot{x}^h, \dot{x}^k$, and contraction of indices i and j, for Eqs. (4.6.7), (4.9.15b), (4.9.16) and (4.9.20) also yield the following results in an *HR*-F_n with non-zero scalar curvature H:

$$\mathfrak{B}_l H^i_{jk} = \lambda_l H^i_{jk} \quad (2.1a); \qquad \mathfrak{B}_l H^i_j = \lambda_l H^i_j; \quad (2.1b)$$

$$\mathfrak{B}_l H_{kh} = \lambda_l H_{kh} \quad (2.2a); \qquad \mathfrak{B}_l H_k = \lambda_l H_k; \quad (2.2b)$$

and

$$\mathfrak{B}_l H = \lambda_l H \tag{2.2c}$$

$\Rightarrow$

$$\lambda_l = (\mathfrak{B}_l H) / H = \mathfrak{B}_l \ln H. \tag{2.2d}$$

Forming the partial derivative of Eq. (2.2b) w.r.t. $\dot{x}^h$, applying the commutation rule of the operators given by Eq. (4.8.3), and using Eqs. (4.6.7) and (4.9.21b), there follows

$$\mathfrak{B}_l H_{kh} - H_i G^i_{khl} = \lambda_l H_{kh} + (\dot{\partial}_h \lambda_l) H_k,$$

or, for Eq. (2.2.a)

$$- H_i G^i_{khl} = (\dot{\partial}_h \lambda_l) H_k. \tag{2.3}$$

Its transvection with $\dot{x}^k$, for Eqs. (4.9.21c) and (4.10.15), establishes the independence of the recurrence vector λ_l on the directional arguments:

$$\dot{\partial}_h \lambda_l = \dot{\partial}_h \mathfrak{B}_l \ln H = 0, \tag{2.4a}$$

by Eq. (2.2d), if $H \neq 0$. Thus, we have the

Theorem 2.1. The recurrence vector λ_l is necessarily a point function in an *HR*-F_n with non-zero scalar curvature *H*.

From Eq. (2.2d) and above theorem, there also follows the:

Corollary 2.1. The covariant derivative of the function ln *H* is also independent of the directional arguments in an *HR*-F_n with non-zero scalar curvature.

Further, the operators $\dot{\partial}_h$ and $\mathfrak{B}_l$ being commutative for a scalar function, the Eq. (2.4a) is also written as:

$$\mathfrak{B}_l\,(\dot{\partial}_h \ln H) = 0, \tag{2.4b}$$

establishing the following result.

Corollary 2.2. The derivative $\dot{\partial}_h(\ln H)$ is also a covariant constant in an an *HR*-F_n with non-zero scalar curvature.

Forming the covariant derivation of Eq. (2.2d) w.r.t. x^m, taking the skew-symmetric part w.r.t. indices *l*, *m*, applying the commutation formula in Eq. (4.7.14) for the scalar *H*, the tensor $A_{l\,m}$ considered in [112] for an H^i_j -recurrent F_n reduces to

$$A_{l\,m} \equiv 2\,\mathfrak{B}_{[l}\ \lambda_{\ m]} = 2\,\mathfrak{B}_{[l}\ \mathfrak{B}_{\ m]} \ln H = -(\dot{\partial}_h \ln H)\ H^h_{l\,m}\,. \tag{2.5}$$

Thus, above identity also holds good in an *HR*-F_n with non-zero scalar curvature. The covariant derivative of above tensor, for Eqs. (2.1a) and (2.4b) gives

$$\mathfrak{B}_k A_{l\,m} \equiv -(\dot{\partial}_h \ln H)\,(\mathfrak{B}_k H^h_{l\,m}) = \lambda_k\ A_{l\,m}\,, \tag{2.6}$$

by Eq. (2.5) itself. This relation provides a shorter proof of the recurrence property of the skew-symmetric tensor $A_{l\,m}$. (Also, cf. [130], Theorem 3.2.).

Note 2.1. The tensor A_{lm} being

$$A_{lm} = \mathfrak{B}_l \lambda_m - \mathfrak{B}_m \lambda_l = \partial_l \lambda_m - \partial_m \lambda_l, \qquad (2.7)$$

by Theorem 2.1, is also a point function in an HR-F_n with non-zero scalar curvature.

2.1. Space of constant Riemannian curvature

A Finsler space F_n $(n > 2)$ of constant Riemannian curvature R is characterized by Eq. (4.11.7). Both the function F and the unit vector l_j being covariant constants vide Eqs. (4.6.7) and (4.6.9), and noting the constant nature of the Riemannian curvature (as per hypothesis), the covariant derivative of Eq. (4.11.7) vanishes. This will also imply the vanishing of covariant derivative of the scalar curvature H as seen in the previous chapter (cf. Eq. (7.2.3)). But, in view of Eq. (2.2d), it causes vanishing of the recurrence vector λ_l producing a contradiction to the very hypothesis of a recurrent space.

Hence, we have the:

Theorem 2.2. A Finsler space F_n $(n > 2)$ of constant Riemannian curvature R cannot be of recurrent curvature.

For example, in 2-dimensional Finsler spaces of constant curvature the tensor H^i_{jk} is also a covariant constant that ultimately causes Eq. (7.2.3). Thus, above theorem also holds good for a F_2.

2.2. An isotropic Finsler space F_n $(n > 2)$

An isotropic Finsler space F_n $(n > 2)$ is characterized by Eqs. (4.11.4) and (4.11.5). Again, the function F and the unit vectors l^i and l_j being covariant constants, the covariant derivatives of these equations, for Eq. (4.8.3), yield

$$\mathfrak{B}_l H^i_{jk} = (2/3)\, F^2 (\delta^i_{[j} - l^i\, l_{[j})\, \dot{\partial}_{k]}\, \mathfrak{B}_l R + 2F \delta^i_{[j}\, l_{k]}\, (\mathfrak{B}_l R), \qquad (2.8)$$

and

$$\mathfrak{B}_l R = (\mathfrak{B}_l H) / F^2. \qquad (2.9)$$

These equations confirm the non-vanishing character of $\mathfrak{B}_l H$. As a result, an isotropic Finsler space F_n $(n > 2)$ may possess recurrent curvature. Thus, for an isotropic $HR\text{-}F_n$ ($n > 2$), Eqs. (2.2c) and (2.9) imply

$$\mathfrak{B}_l\, R = \lambda_l\, (H / F^2) = \lambda_l\, R, \tag{2.10}$$

by Eq. (4.11.5). This establishes the:

Theorem 2.3. The Riemannian curvature of an isotropic $HR\text{-}F_n$ ($n > 2$) is recurrent.

2.3. *WR-F$_n$* ($n > 2$)

Definition 2.1. A Finsler space F_n , $n > 2$, is called a *projectively recurrent* Finsler space (briefly denoted as $WR\text{-}F_n$) if its projective curvature tensor W^i_{jkh} satisfies an equation analogous to Eq. (1.1):

$$\mathfrak{B}_l\, W^i_{jkh} = \lambda_l\, W^i_{jkh}. \tag{2.11}$$

For Eq. (4.6.7), it may be seen from Eqs. (5.2.13c), (1.1), (2.1) and (2.2) that the projective tensor W^i_{jk} is also recurrent in an $HR\text{-}F_n$:

$$\mathfrak{B}_l\, W^i_{jk} = \lambda_l\, W^i_{jk}. \tag{2.12}$$

Its transvection with $\dot{x}^k$, for Eqs. (4.6.7) and (5.2.17) also yields

$$\mathfrak{B}_l\, W^i_j = \lambda_l\, W^i_j. \tag{2.13}$$

As in § 5 of Chapter 7, the directional derivative of Eq. (2.12), use of commutation rule in Eq. (4.8.3) and Eqs. (5.2.14) and (2.4a), yields

$$\mathfrak{B}_l\, W^i_{jkh} + W^m_{jk}\, G^i_{hlm} - W^i_{mk}\, G^m_{jhl} - W^i_{jm}\, G^m_{khl} = \lambda_l\, W^i_{jkh}. \tag{2.14}$$

It reduces to Eq. (2.11), if there holds

$$W^m_{jk}\, G^i_{hlm} - W^i_{mk}\, G^m_{jhl} - W^i_{jm}\, G^m_{khl} = 0, \tag{2.15}$$

which is same as Eq. (7.5.5). Thus, in general, an $HR\text{-}F_n$, $n > 2$, need not be projectively recurrent unless there holds Eq. (2.15). However, the tensor G^i_{jkh} vanishes identically in an affinely connected space, above condition gets automatically satisfied. Hence, we have

Theorem 2.4. An $HR\text{-}F_n$, $n > 2$, becomes projectively recurrent if it is an affinely connected space.

As in § 2 of Chapter 7, an application of the commutation formula (4.8.3) to the curvature tensor H^{i}_{jk} satisfying Eq. (2.1a) in an $HR\text{-}F_n$, also yields the identity:

$$H^{m}_{jk}\, G^{i}_{hlm} + 2H^{i}_{m\,[j}\, G^{m}_{k]hl} = 0, \qquad (2.16)$$

for Eqs. (4.9.15a), (1.1) and (2.4a). This is same as Eq. (7.2.6). As in Chapter 7, its contraction w.r.t. indices i and j also yields Eq. (7.2.7). Further, as in § 6 of Chapter 7, plugging in for the projective tensors from Eq. (5.2.13c), and use of Eqs. (4.10.15), (7.2.7) and (2.16), yields the alternate form of Eq. (2.15):

$$G^{p}_{lm\,[j}\,\{\dot{x}^{i}\, H^{r}_{k]\,pr} + \delta^{i}_{k]}(\dot{x}^{r}\, H_{rp})/(n-1)\} = 0, \qquad (2.17)$$

which is same as Eq. (7.6.2).

§ 3. *HR-F$_n$* with special reference to Lie derivation

Analogous to the terminology used for symmetric and recurrent spaces, we propose the following definition.

Definition 3.1. If the Lie derivative of a geometric object Ω w.r.t. the infinitesimal transformation given by Eq. (5.4.1) is zero, it is called a *Lie-invariant*:

$$\pounds\, \Omega = 0; \qquad (3.1)$$

alternately, if $\pounds\, \Omega$ is proportional to the object itself for some non-null scalar function μ (say), it is called *Lie-recurrent*:

$$\pounds\, \Omega = \mu\, \Omega, \qquad \mu \neq 0. \qquad (3.2)$$

Lie derivative of the curvature tensor H^{i}_{jkh}, found by Eq. (6.2.16b), for Eq. (1.1), reads as:

$$\pounds\, H^{i}_{jkh} = L\, H^{i}_{jkh} - H^{l}_{jkh}\, \mathfrak{B}_l\, v^{i} - 2(\mathfrak{B}_{[j}\, v^{l})\, H^{i}_{k]lh}$$
$$+ H^{i}_{jkl}\, \mathfrak{B}_h\, v^{l} + (\dot{\partial}_l\, H^{i}_{jkh})\,(\mathfrak{B}_m\, v^{l})\,\dot{x}^{m}. \qquad (3.3)$$

where

$$L \equiv v^{l}\lambda_l, \qquad (3.4)$$

is a non-zero scalar point function unless the vector-fields v^{l} and λ_l are orthogonal to each other. Transvection of Eq. (3.3) with $\dot{x}^k\,\dot{x}^h$ and contraction of indices i and j, for Eqs. (4.9.15b), (4.9.16), (4.9.20c), (4.9.21), (4.9.22) and (6.2.9b) determines the Lie-recurrence of the scalar curvature H:

$$£\, H = L\, H + (\dot{\partial}_l\, H)\,(\mathfrak{B}_m\, v^l)\,\dot{x}^m, \tag{3.5}$$

when $L \neq 0$. Thus, we have the:

Theorem 3.1. The scalar curvature is, in general, neither Lie-recurrent nor Lie-invariant in an HR-F_n.

Note 3.1. Equation (3.5) can be directly derived from the formula (6.2.3d), where substitutions are made from Eqs. (2.2c) and (3.4).

Analogous to Eq. (7.3.3) holding in an S-F_n, the commutation formula vide Eq. (4.7.14) applied to the curvature tensor H^i_{jkh}, for Eq. (1.1), yields

$$A_{lm}\, H^i_{jkh} = H^i_{lmr}\, H^r_{jkh} + 2H^r_{lm[j}\, H^i_{k]rh} - H^r_{lmh}\, H^i_{jkr} - (\dot{\partial}_r\, H^i_{jkh})\, H^r_{lm}, \tag{3.6}$$

where the skew-symmetric tensor A_{lm} is given by Eq. (2.5). Repeating the process of transvection by $\dot{x}^k\,\dot{x}^h$, contraction w.r.t. the indices i and j in Eq. (3.6) and using Eq. (4.9.22), the Eq. (2.5) can be again derived.

The tensor A_{lm} plays an important role in our discussion. Let us assume a conjecture that there may exist a skew-symmetric tensor f^{jk} satisfying

$$\mathfrak{B}_h\, v^i = f^{jk}\, H^i_{jkh} \equiv J^i_h \tag{3.7}$$

in the space HR-F_n. Transvection of Eq. (3.6) by f^{lm}, for above assumption gives

$$f^{lm} A_{lm}\, H^i_{jkh} = H^r_{jkh}\, J^i_r + 2J^{\ r}_{[j}\, H^i_{k]rh} - H^i_{jkr}\, J^r_h - (\dot{\partial}_r\, H^i_{jkh})\, J^r_p\, \dot{x}^p, \tag{3.8}$$

which, on comparison with Eq. (3.3), yields

$$£\, H^i_{jkh} = (L - f^{lm} A_{lm})\, H^i_{jkh}. \tag{3.9}$$

This establishes the

Theorem 3.2. The curvature tensor H^{i}_{jkh} becomes Lie-invariant in *HR*-F_n if and only if there holds the condition:

$$L = f^{lm} A_{lm} \tag{3.10}$$

for arbitrary H^{i}_{jkh}.

3.1. λ_m a gradient vector

Particularly, if the recurrence vector λ_m is a gradient vector satisfying

$$\lambda_m = \partial_m \lambda = \mathfrak{B}_m \lambda , \tag{3.11}$$

vanishing of the tensor A_{lm} may be easily verified from Eq. (2.7):

$$A_{lm} = 0. \tag{3.12}$$

Conversely, when Eq. (3.12) holds, Eq. (2.7) yields $\partial_l \lambda_m = \partial_m \lambda_l$ implying exact form of differential equation $\lambda_m dx^m = 0$ confirming the gradient form of vector λ_m. Thus, we have the:

Lemma 3.1. The recurrence vector λ_m of an *HR*-F_n is a gradient vector if and only if the tensor A_{lm} vanishes.

Hence, Eq. (3.9) reduces to

$$\pounds H^{i}_{jkh} = L H^{i}_{jkh}. \tag{3.13}$$

Thus, we may propose a theorem [130], Theo. 2.2 [31)]:

Theorem 3.3. Any two of the following conditions holding in an *HR*-F_n imply the remaining one:

(i) the recurrence vector λ_m is a gradient vector,

(ii) the curvature tensor H^{i}_{jkh} is a Lie-invariant,

(iii) the function L vanishes.

Proof. When the first two conditions hold satisfying Eq. (3.12) and making LHS of Eq. (3.9) zero gives rise to the third condition automatically as $H^{i}_{jkh} \neq 0$ in *HR*-F_n . Similarly, the first and the last conditions satisfying Eq. (3.12) also make the LHS of Eq. (3.9) zero and establish the second condition.

Lastly, when the second and third conditions hold, it follows from Eqs. (3.9) and (3.10):

$$f^{lm} A_{lm} = 0, \tag{3.14}$$

or, by Eq. (2.5)

$$-(\dot{\partial}_h \ln H)(f^{lm} H^h_{lm}) = 0 \quad \Rightarrow \quad \dot{\partial}_h \ln H = 0, \tag{3.15}$$

proving that H is a point function. Consequently, Eq. (2.5) yields back Eq. (3.12). Thus, the gradient property of the vector λ_l is established in view of Eq. (2.7) and theory of ordinary differential equations [122]. //

3.2. Lie invariance of curvature tensor

Multiplying Eqs. (3.3) and (3.6) by A_{lm} and L respectively and adding the resulting equations, there follows the identity

$$\begin{aligned} A_{lm} \pounds H^i_{jkh} = {} & H^r_{jkh}(LH^i_{lmr} - A_{lm}J^i_r) - H^i_{rkh}(LH^r_{lmj} - A_{lm}J^r_j) \\ & - H^i_{jrh}(LH^r_{lmk} - A_{lm}J^r_k) - H^i_{jkr}(LH^r_{lmh} - A_{lm}J^r_h) \\ & - (\dot{\partial}_r H^i_{jkh})(LH^r_{lms} - A_{lm}J^r_s)\dot{x}^s. \end{aligned} \tag{3.16}$$

Thus, vanishing of the Lie derivative of curvature tensor is implied by the relation

$$L H^i_{jkh} = A_{jk} J^i_h = A_{jk} \mathfrak{B}_h v^i. \tag{3.17}$$

On the other hand, if H^i_{jkh} is Lie-invariant and so is the scalar curvature H for Eq. (6.8.10), transvection of Eq. (3.16) with $\dot{x}^k \dot{x}^h$ and contraction of indices i and j, for Eqs. (4.9.15b), (4.9.16), (4.9.20), (4.9.21) and (6.2.9b) yield

$$(H_r + H_{kr}\dot{x}^k)(L H^r_{lms} - A_{lm}J^r_s)\dot{x}^s = 0,$$

or, by Eq. (4.9.22)

$$(\dot{\partial}_r H)(L H^r_{lms} - A_{lm}J^r_s)\dot{x}^s = 0, \tag{3.18}$$

implying the condition vide Eq. (3.17) for linearly independent components $\dot{x}^i$'s and non-zero character of $\dot{\partial}_r H$, in general.

Thus, we have the

Theorem 3.4. The curvature tensor of an *HR*-F_n is Lie-invariant if and only if it admits Eq. (3.17).

3.3. $\pounds \lambda_i$ a parallel vector-field

The concept of parallelism of vectors given by Levi-Civita for Riemannian spaces is analogously extended to Finslerian geometry [189], p. 67.

Definition 3.2. A vector will be called *parallel* if it possesses vanishing covariant derivative.

Thus, a parallel vector-field $\pounds \lambda_j$ satisfies

$$\mathfrak{B}_i \pounds \lambda_j = 0. \tag{3.19}$$

Hence, the commutation formula (6.6.7), applied to the vector λ_j and independence of the vector on $\dot{x}^i$'s, reduces to

$$\pounds \mathfrak{B}_i \lambda_j = -\lambda_k \pounds G^k_{i\,j}. \tag{3.20}$$

Its skew-symmetric part w.r.t. the indices i and j, for symmetry of $G^k_{i\,j}$ in its lower indices and Eq. (1.2),

$$\pounds A_{ij} \equiv 2 \pounds \mathfrak{B}_{[i} \lambda_{j]} = 2 \mathfrak{B}_{[i} \pounds \lambda_{j]}, \tag{3.21}$$

by Eq. (6.6.7). Further, by Eq. (6.2.12c), the vector

$$\pounds \lambda_j = v^a (\mathfrak{B}_j \lambda_a) + \lambda_a (\mathfrak{B}_j v^a) = v^a (\partial_j \lambda_a) + \lambda_a (\partial_j v^a), \tag{3.22}$$

is a point function for both the vectors λ_j and v^i being so. Hence, the vector in Eq. (3.21) becomes

$$\pounds A_{ij} = 2 \mathfrak{B}_{[i} \pounds \lambda_{j]} = (\partial_i \pounds \lambda_j - \partial_j \pounds \lambda_i), \tag{3.23}$$

and is a point function. It vanishes, for Eq. (3.19), making the tensor A_{ij} Lie-invariant in an *HR*-F_n with a parallel vector field:

$$\pounds A_{ij} = 0. \tag{3.24}$$

Thus, we have the:

Theorem 3.5. The tensor field $A_{l\,m}$ is Lie-invariant in an HR-F_n if its recurrence vector is a parallel vector-field.

Conversely, if there holds Eq. (3.24), it follows from Eq. (3.23) that the vector-field £ λ_j is a gradient vector but not a parallel vector-field.

§ 4. H^{i}_{jk} -recurrent space

The possibilities for H^{i}_{jk} -recurrent space, satisfying Eq. (2.1a), to have recurrent curvature were explored by the author [110]. Like § 2 of Chapter 7, application of the commutation formula in Eq. (4.8.3) to the tensor H^{i}_{jk} , and use of Eqs. (4.9.15a), (2.1a) and (2.4a), yields

$$(\mathfrak{B}_l - \lambda_l) H^{i}_{jkh} = - H^{r}_{jk}\, G^{i}_{rlh} - 2\, H^{i}_{r[j}\, G^{r}_{k]lh}. \tag{4.1}$$

Note 4.1. Since Eq. (2.2c) also holds in an H^{i}_{jk} -recurrent space, the discussion of § 2 leading to Eqs. (2.5) and (2.6) is also valid in this space.

4.1. To explore if the H^{i}_{jk} -recurrent space F_n were of recurrent curvature the LHS of Eq. (4.1) must vanish requiring

$$H^{r}_{jk}\, G^{i}_{rlh} + 2\, H^{i}_{r[j}\, G^{r}_{k]lh} = 0. \tag{4.2}$$

which is same as Eq. (7.2.6). This identity, resulting from the commutation formula (4.8.3) applied to the tensor H^{i}_{jk} confirms commutation of the operators of covariant and directional derivations and represents a necessary condition for an H^{i}_{jk} -recurrent space F_n to become an HR-F_n. Conversely, when this identity holds in an H^{i}_{jk} -recurrent space F_n , the LHS of Eq. (4.1) vanishes. Hence, the condition vide Eq. (4.2) is also sufficient for an H^{i}_{jk} -recurrent space F_n to become an HR-F_n. Thus, we have the:

Theorem 4.1. An H^{i}_{jk} -recurrent space F_n becomes of recurrent curvature if and only if the processes of covariant and directional derivation are commutative over each other leading to the identity (4.2).

Since the tensor G^i_{jkh} vanishes identically in Berwald space (cf. Eq. (3.10.4)), the Eq. (4.2) gets satisfied in the space. Hence, we have the:

Corollary 4.1. An H^i_{jk}-recurrent space F_n becomes an *HR-*F_n if it is a Berwald space.

However, its converse statement need not hold, i.e. an H^i_{jk}-recurrent space becoming an *HR-*F_n need not be a Berwald space because, the Eq. (4.2), on contraction w.r.t. indices *i* and *j*, only yields

$$H_r\, G^r_{klh} = 0, \tag{4.3}$$

and not necessarily Eq. (3.10.4).

§ 5. H^i_j-recurrent space F_n

An H^i_j-recurrent space is characterized by Eq. (2.1b). As in the previous section, directional derivative of Eq. (2.1b) w.r.t. $\dot{x}^k$, for the commutation formula (4.8.3) applied to the tensor H^i_j, there results

$$(\mathfrak{B}_l - \lambda_l)\dot{\partial}_k H^i_j = H^i_r\, G^r_{jkl} - H^r_j\, G^i_{rkl}. \tag{5.1}$$

Its skew-symmetric part w.r.t. the indices *j* and *k*, and multiplication by 1/3, for Eq. (4.9.17) and symmetry of the tensor G^r_{jkl} in its lower indices, yields

$$(\mathfrak{B}_l - \lambda_l) H^i_{jk} = -(2/3) H^r_{[j}\, G^i_{k]lr}. \tag{5.2}$$

5.1. To explore if the H^i_j-recurrent space becomes H^i_{jk}-recurrent the LHS of Eq. (5.2) must vanish requiring

$$H^r_{[j}\, G^i_{k]lr} = 0. \tag{5.3}$$

Thus, analogous to Theorem 4.1, there holds the:

Theorem 5.1. An H^i_j-recurrent space F_n becomes H^i_{jk}-recurrent space if and only if the processes of covariant and directional derivation commute over each other leading to the identity (5.3).

As mentioned in Note 4.1, the Eqs. (2.5) and (2.6) also hold good in an $H^{\,i}_{j}$ -recurrent space F_n.

5.2. When both the conditions vide Eqs. (4.2) and (5.3) hold simultaneously an $H^{\,i}_{j}$ -recurrent space F_n becomes simultaneously an $H^{\,i}_{jk}$-recurrent as well as $HR\text{-}F_n$.

Thus, Theorem 5.5 of [110] was amended by the author in [112]:

Theorem 5.2. An $H^{\,i}_{j}$ -recurrent space F_n becomes simultaneously an $H^{\,i}_{jk}$-recurrent as well as $HR\text{-}F_n$ if there hold the conditions vide Eqs. (4.2) and (5.3).

As seen earlier (cf. Eq. (3.10.4)) the tensor G^{i}_{jkh} vanishes identically in a Berwald space so the conditions vide Eqs. (4.2) and (5.3) get obviously satisfied. Thus, analogous to the Corollary 4.1, there also hold the:

Corollary 5.1. An $H^{\,i}_{j}$ -recurrent space F_n becomes an $HR\text{-}F_n$ if it is a Berwald space.

§ 6. *HR-F*$_n$ with infinitesimal transformations of special form

6.1. Special concircular vector field

Extending the discussion of Sub-section 9.1 of Chapter 5 to an $HR\text{-}F_n$, where hold the Eqs. (1.1) and (2.2), the Eq. (5.9.16) reduces to

$$(v^h\lambda_l + \rho\,\delta^h_l)\{H^{i}_{jkh} - 2\,\delta^{i}_{[j}\,H_{k]h}/(n-1)\} = 0, \quad (6.1)$$

or, by Eq. (5.9.15b)

$$H^{i}_{jkh} = 2\,\delta^{i}_{[j}\,H_{k]h}/(n-1). \quad (6.2)$$

On contraction w.r.t. the indices i and h, for Eq. (4.10.3), it yields

$$H_{kj} - H_{jk} = (H_{kj} - H_{jk})/(n-1) \quad \Rightarrow \quad H_{kj} = H_{jk}, \quad (6.3)$$

if $n > 2$. Thus, analogous to Theorem 3.1 of [135], there holds:

Theorem 6.1. The contracted curvature tensor H_{jk} is symmetric in an *HR-F*$_n$ $(n > 2)$ admitting a special concircular infinitesimal transformation.

As a consequence of this theorem, Eq. (4.9.21a) may be alternately written as

$$H_{jk}\,\dot{x}^j = H_k\,. \tag{6.4}$$

Also, transvection of Eq. (6.2) with $\dot{x}^h$, for Eqs. (4.9.15b) and (4.9.21a), determines

$$H^i_{jk} = 2\,\delta^i_{[j}\,H_{k]}\,/(n-1)\,. \tag{6.5}$$

Hence, putting from the Eqs. (6.3) - (6.5) in Eq. (5.2.13c), we notice the vanishing of projective tensor W^i_{jk}:

$$W^i_{jk} = 0\,. \tag{6.6}$$

This, on derivation w.r.t. $\dot{x}^h$, for Eq. (5.2.14), yields

$$W^i_{jkh} = 0, \tag{6.7}$$

and, on transvection with $\dot{x}^k$, for Eq. (5.2.17), also yields

$$W^i_j = 0\,. \tag{6.8}$$

Thus, the Corollary 3.1 of [135] also holds in the space under consideration.

Corollary 6.1. The space *HR-F*$_n$ admitting a special concircular infinitesimal transformation is projectively flat.

6.2. A relation for the tensor H^i_{jk}

In the following we deduce a relation connecting the tensor H^i_{jk} in terms of the recurrence vector λ_i. The covariant derivation of Eq. (6.5), for Eq. (2.2b), yields

$$\mathfrak{B}_h\,H^i_{jk} = \lambda_h\,(\delta^i_j\,H_k - \delta^i_k\,H_j)/(n-1)\,. \tag{6.9}$$

Interchanging the indices *j*, *k*, *h* cyclically in Eq. (6.9) and adding the equations so obtained, there results

$$2\{\lambda_h \delta^i_{[j} H_{k]} + \lambda_j \delta^i_{[k} H_{h]} + \lambda_k \delta^i_{[h} H_{j]}\} = 0,$$

where Bianchi identity (4.10.13) is also used. On contraction w.r.t. the indices *i* and *j*, it yields

$$\lambda_k H_h = \lambda_h H_k. \tag{6.10}$$

Finally, its transvection with $\dot{x}^h$, for Eq. (4.9.21c), determines

$$H_k = (n-1)(H/\lambda)\lambda_k, \tag{6.11}$$

where we have put $\lambda \equiv \lambda_h \dot{x}^h$. Accordingly, Eq. (6.5) reduces to

$$H^i_{jk} = 2(H/\lambda)\delta^i_{[j} \lambda_{k]}. \tag{6.12}$$

6.3. Special c.i.t.'s in a recurrent non-Riemannian space

Let A_n^* be a non-Riemannian space of recurrent curvature admitting a special concircular infinitesimal transformation so that there hold equations analogous to (5.9.1c), (5.9.11b), (5.9.14), (1.1), (2.2) and (6.3) making contracted curvature tensor H_{kh} symmetric. The covariant derivation of Eq. (5.9.14) w.r.t. x^j, for Eqs. (5.9.1c), (2.2) and (5.9.14) itself, gives

$$\mathfrak{B}_j \rho_k = \lambda_j \rho_k + \rho H_{kj}/(1-n). \tag{6.13}$$

Both the tensors $\mathfrak{B}_j \rho_k$ and H_{kj} being symmetric vide Eqs, (5.9.11b) and (6.3), the skew-symmetric part of above equation w.r.t. indices *j* and *k*, yields

$$\lambda_j \rho_k - \lambda_k \rho_j = 0, \quad \text{i.e.} \quad \lambda_j \rho_k = \lambda_k \rho_j, \tag{6.14}$$

suggesting a proportionality relation between the two vectors:

$$\rho_k = \alpha \lambda_k, \tag{6.15}$$

for some arbitrary non-null scalar point function α for both ρ_k and λ_k being so).

The Lie derivative of the curvature tensor given by Eq. (3.3), for Eqs. (5.9.1c), (1.1) and (3.4), becomes

$$\pounds H^{i}_{jkh} = (L + 2\rho)\, H^{i}_{jkh}. \tag{6.16}$$

Depending upon the nature of scalar function $L + 2\rho$, above equation gives rise to different cases.

Case 6.1. For non-zero choice of $L + 2\rho$, the Eq. (6.16) states a recurrent property of the curvature tensor under Lie derivation and the space under consideration is called a *Lie-recurrent space*.

Case 6.2. For vanishing case:

$$L + 2\rho = 0, \tag{6.17}$$

Eq. (6.16) shows the Lie-invariant property of the curvature tensor and so will be the contracted curvature tensor. Hence, the transformation defines a curvature collineation (cf. § 10 of Chapter 6). Also, the Lie derivative of the vector ρ_k obtained by Eq. (6.2.12c), for Eqs. (5.9.11a), for Eq. (5.9.1c), becomes zero:

$$\pounds \rho_k = v^j (\mathfrak{B}_j \rho_k) + \rho_j (\mathfrak{B}_k v^j) = v^j \mathfrak{B}_j \rho_k + \rho \rho_k = 0, \tag{6.18a}$$

or, for Eqs. (3.4) and (6.17)

$$v^j \{\mathfrak{B}_j \rho_k - \lambda_j \rho_k / 2\} = 0. \tag{6.18b}$$

Thus, we have the

Theorem 6.2. A non-Riemannian space admitting a special concircular infinitesimal transformation is either Lie-recurrent or satisfies the following consequences:

(i) it admits a curvature collineation; and
(ii) The vector field ρ_k is Lie invariant satisfying the identities (6.18).

6.4. Contra infinitesimal transformations in *HR*-F_n

For infinitesimal transformations generated by a vector field of contra form, the Eq. (3.3) assumes the form of Eq. (3.13). Hence, analogous to Cases 6.1. and 6.2 there arise the following situations.

Case 6.3. For non-zero L, the Eq. (3.13) states Lie-recurrent property of the curvature tensor. The contracted curvature tensors and other associated tensors will also be of same nature:

$$(£-L)H_{kh} = (£-L)H^{i}_{jk} = (£-L)H^{i}_{j} = (£-L)H_{k} = (£-L)H = 0. \tag{6.19}$$

Case 6.4. When $L = 0$, it follows from Eq. (3.4) that the vector fields v^{i} and λ_i are orthogonal to each other and the curvature tensor along with its contracted tensors and associated tensors are Lie invariants - hence leading to curvature collineation:

$$£\,H_{kh} = £\,H^{i}_{jk} = £\,H^{i}_{j} = £\,H_{k} = £\,H = 0. \tag{6.20}$$

6.5. Recurrent vector-field

Continuing the discussion of § 10 of Chapter 5, the Bianchi identities vide Eqs. (4.10.5) and (4.10.13) in an *HR-F_n* admitting an infinitesimal transformation generated by a recurrent vector-field, for Eq. (1.1), reduce to

$$\lambda_{[j} H^{i}_{kh]m} + G^{i}_{lm[j} H^{l}_{kh]} = 0, \tag{6.21}$$

and

$$\lambda_{[j} H^{i}_{kh]} = 0. \tag{6.22}$$

On transvection with the vector v^m, for Eq. (5.10.2), above Eq. (6.21) further reduces to

$$\lambda_{[j} H^{i}_{kh]m} v^{m} + v^{i} \dot{\partial}_{l} \mu_{[j} H^{l}_{kh]} = 0. \tag{6.23}$$

§ 7. Affine motions in an *HR-F_n*

Affine motions in a general Finsler space F_n are studied in Chapter 6, while those for a symmetric and projectively symmetric Finsler spaces were discussed in Chapter 7. Presently, we consider such motions in recurrent Finsler spaces. Thus, all Eqs. (6.8.1) - (6.8.3), (6.8.5) - (6.8.7), (6.8.10), (6.8.12) and (1.1), (2.1) - (2.2) are consistent in an *HR-F_n* admitting an affine motion. This study was carried out by Sinha [213] and the present author with Meher [130]. An *HR-F_n* admitting an affine motion has been denoted by *AHR-F_n* [130].

Forming Lie derivative of Eq. (1.1), applying the commutation rule vide Eq. (6.6.7) and putting from Eqs. (6.8.1), (6.8.2) and (6.8.5), there follows the result

$$(\pounds \lambda_m) H^i_{jkh} = 0 \quad \Rightarrow \quad \pounds \lambda_m = 0, \tag{7.1}$$

for arbitrary choice of tensor H^i_{jkh}. This result was derived by Sinha [213] in an *AHR-F*$_n$. As per Eq. (6.8.10), the Lie derivative of the scalar curvature H vanishes in an *AHR-F*$_n$. This reduces the Eq. (3.5) determining the function L:

$$L = -(\dot{\partial}_l \ln H)(\mathfrak{B}_m v^l)\dot{x}^m. \tag{7.2}$$

Hence, Eq. (3.10) reduces to

$$L = f^{lm} A_{lm} = -(\dot{\partial}_l \ln H)(\mathfrak{B}_m v^l)\dot{x}^m. \tag{7.3}$$

These observations lead to the

Theorem 7.2. There exists a skew-symmetric tensor field f^{jk} satisfying Eqs. (3.7) and (7.3) in an *AHR-F*$_n$.

Writing the Lie derivative of the recurrence vector λ_k by the formula vide Eq. (6.2.12c):

$$\pounds \lambda_k = v^j(\mathfrak{B}_j \lambda_k) + \lambda_j(\mathfrak{B}_k v^j) = v^j A_{jk} + \mathfrak{B}_k L, \tag{7.4}$$

by Eq. (2.7); and reducing it for an *AHR-F*$_n$ admitting Eq. (7.1), there follows

$$v^j A_{jk} + \mathfrak{B}_k L = 0. \tag{7.5}$$

Introducing a vector-field η^k satisfying

$$\eta^k \mathfrak{B}_k L = L, \tag{7.6}$$

a transvection of Eq. (7.5) with η^k yields the relation

$$v^j \eta^k A_{jk} = -L. \tag{7.7}$$

Hence, transvection of Eq. (3.17), holding in an *AHR-F*$_n$, by $v^j \eta^k$, gives

$$\mathfrak{B}_h v^i = -v^j \eta^k H^i_{jkh} = v^k \eta^j H^i_{jkh}, \tag{7.8}$$

for skew-symmetric properties of H^i_{jkh}. These observation lead to the:

Theorem 7.3. If an $HR\text{-}F_n$ admits an affine motion there exists a vector field η^k satisfying Eq. (7.8).

In view of Note 6.8.1, we have

$$\text{£}\,\mathfrak{B}_l\,\lambda_m = \mathfrak{B}_l\,\text{£}\,\lambda_m = 0, \tag{7.9}$$

by Eq. (7.1) in an $AHR\text{-}F_n$. Hence, Eq. (2.5) again establishes Eq. (3.24) and analogous to Theorem 3.5, we have

Theorem 7.4. The tensor field A_{lm} is also Lie-invariant in an $AHR\text{-}F_n$.

§ 8. Condition for $HR\text{-}F_n$ to become $AHR\text{-}F_n$

Vanishing of Lie derivative of the connection parameters, exhibited by Eq. (6.8.1), implies Lie-invariance of the curvature tensor $H^{\,i}_{j\,k\,h}$ and the recurrence vector λ_m (cf. Eqs. (6.8.5) and (7.1)). Thus, Eqs. (6.8.5) and (7.1) are the necessary consequences of Eq. (6.8.1). Further, in Theorem 3.4, Eq. (3.17) has been seen as a necessary and sufficient condition for Lie-invariance of the curvature tensor. In the following, we examine the converse problem if Eqs. (6.8.5) and (7.1) may be sufficient to cause Eq. (6.8.1) in an $HR\text{-}F_n$?

Differentiating Eq. (3.17) covariantly w.r.t. x^l and putting from Eqs. (1.1), (2.6) and (3.17) itself, there follows

$$(\mathfrak{B}_l\ L)\ H^{\,i}_{j\,k\,h} = A_{j\,k}\,\mathfrak{B}_l\,\mathfrak{B}_h\,v^i\,. \tag{8.1}$$

Its transvection by $v^j\,\eta^k$ and use of Eq. (7.7) yields

$$\mathfrak{B}_l\,\mathfrak{B}_h\,v^i = -\,v^j\,\eta^k\,(\mathfrak{B}_l\ \ln L)\,H^{\,i}_{j\,k\,h}, \tag{8. 2a}$$

or, by Eq. (7.8)

$$\mathfrak{B}_l\,\mathfrak{B}_h\,v^i = (\mathfrak{B}_l\ \ln L)\,\mathfrak{B}_h\,v^{\,i}. \tag{8. 2b}$$

Thus, the covariant derivatives of the vector $v^{\,i}$ generating the infinitesimal transformation given by Eq. (5.4.1) in an $HR\text{-}F_n$ that admits the Eqs. (6.8.5) and (7.1) are found by Eqs. (7.8) and (8. 2b).

Further, a transvection of Eq. (3.17) by v^j, for Eq. (7.5), yields

$$v^j H^i_{jkh} = -(\mathfrak{B}_k \ln L)\, \mathfrak{B}_h v^i \,. \tag{8.3}$$

Putting from Eqs. (8.2b) and (8.3) in Eq. (6.3.7), the Lie derivative of the connection parameters becomes

$$\pounds G^i_{jk} = G^i_{jkh}\, (\mathfrak{B}_l\, v^h)\, \dot{x}^l \,, \tag{8.4}$$

which does not vanish , in general, unless the RHS expression in above equation is zero. Thus, Eq. (6.8.5) and (7.1) do not establish the existence of an affine motion in an *HR-F*$_n$. For independence of the recurrence vector λ_l on the directional coordinates, the Theorem 4.1 of [130] is amended as

Theorem 8.1. An *HR-F*$_n$ admits an affine motion if there holds the condition

$$G^i_{jkh}\, (\mathfrak{B}_l\, v^h)\, \dot{x}^l = 0. \tag{8.5}$$

Corollary 8.1. An *HR-F*$_n$ admits affine motions generated by contra, concurrent and special concircular vector fields v^i.

Proof. The first case is evident by the definition of a contra affine motion given by Eq. (5.9.1a). Also, for other two cases, the LHS expression of Eq. (8.5), for Eq. (4.10.15), reduces to

$$G^i_{jkh}\, \rho\, \delta^h_l\, \dot{x}^l = \rho\, G^i_{jkh}\, \dot{x}^h = 0,$$

proving the result. //

In the following, a relation connecting the tensor field A_{jk} with the recurrence vector λ_i and the function L is derived. Differentiating Eq. (7.5) covariantly w.r.t. x^h, and noting Eq. (2.6), there results

$$A_{jk}\, \mathfrak{B}_h\, v^j + v^j \lambda_h\, A_{jk} + \mathfrak{B}_h \mathfrak{B}_k L = 0.$$

Forming its skew-symmetric part w.r.t. the indices k and h, we get

$$2\, \{A_{j[k}\, \mathfrak{B}_{h]}\, v^j + v^j A_{j[k}\, \lambda_{h]} + \mathfrak{B}_{[h} \mathfrak{B}_{k]} L\} = 0.$$

The function L, defined by Eq. (3.4), is a point function as both the vector fields v^l and λ_l are independent of $\dot{x}^i$'s. Hence, the last term in above equation vanishes identically for the commutation formula vide Eq. (4.7.14) applied to the function L reducing above equation to

$$2\,\{A_{j[k}\,\mathfrak{B}_{h]}\,v^j + v^j A_{j[k}\,\lambda_{h]}\} = 0. \tag{8.6}$$

On the other hand, the tensor-field A_{kh} is a Lie-invariant point function in an *AHR-*F_n in view of Note 2.1 and Theo. 7.4. Writing its Lie derivative by Eq. (6.2.12c), we thus have

$$\pounds\, A_{kh} \equiv v^j\, \mathfrak{B}_j A_{kh} + A_{jh}\,\mathfrak{B}_k\, v^j + A_{kj}\,\mathfrak{B}_h v^j = 0,$$

or, by Eqs. (2.5), (2.6), (3.4) and skew-symmetry of the tensor A_{kh}

$$L\,A_{kh} = 2A_{j\,[k}\,\mathfrak{B}_{h]}\,v^l, \tag{8.7}$$

or,

$$L\,A_{kh} = -2\,v^j A_{j\,[k}\,\lambda_{h]}, \qquad \text{by Eq. (8.6)}$$

or,

$$A_{kh} = 2\mathfrak{B}_{[k}\,(\ln L)\,\lambda_{h]}, \qquad \text{by Eq. (7.5).}$$

Note 8.1. The last relation is in analogy with its Riemannian counterpart [222], Eq. (3.6).

§ 9. Recurrent affine motion in *HR-*F_n

The vector-field v^i generating an infinitesimal transformation vide Eq. (5.4.1) and satisfying Eq. (5.9.1d) has been called a recurrent vector field. The corresponding affine motion is called a *recurrent affine motion*. The second order covariant derivatives of this vector v^i were evaluated in a general Finsler space F_n in § 9 of Chapter 5. This discussion was continued by the author [112] and author with Meher [128] for spaces of recurrent curvature. However, it was called there as special recurrent affine motion and the underlying space was called a special Finsler space of recurrent curvature, briefly denoted as *SHR-*F_n. Thus, Eqs. (6.8.1) and (6.8.2), for Eq. (5.9.5) and (4.10.15), reduce to

$$v^i\,\mu_{kh} + (H^i_{jkh} + G^i_{jkh}\,\mu_r\,\dot{x}^r)\,v^j = 0, \tag{9.1}$$

and

$$v^i\,\dot{x}^h\,\mu_{kh} + v^j H^i_{jk} = 0. \tag{9.2}$$

Covariant derivation of above equation, for Eqs. (4.6.7), (5.9.1d), (1.1) and (2.1a), yields

$$(\mathfrak{B}_l\,\mu_{kh} + \mu_l\,\mu_{kh})\,v^i\,\dot{x}^h + (\mu_l + \lambda_l)\,v^j H^i_{jk} = 0. \tag{9.3}$$

On the other hand, transvection of Eq. (9.2) with v^k, reduces the second term to zero for skew-symmetric properties of the tensor H^i_{jk}, giving

$$v^i\,(v^k\,\dot{x}^h\,\mu_{kh}) = 0 \;\Rightarrow\; v^k\,\dot{x}^h\,\mu_{kh} = 0, \qquad (9.4)$$

for arbitrary v^i. Further, its covariant derivation, for Eqs. (4.6.7) and (5.9.1d), yields

$$(\mathfrak{B}_l\,\mu_{kh} + \mu_l\,\mu_{kh})\,v^k\,\dot{x}^h = 0. \qquad (9.5)$$

Contraction of Eq. (9.3) for indices i and k and substitution from Eqs. (4.9.20b) and (9.5), yields

$$(\mu_l + \lambda_l)\,v^j H_j = 0. \qquad (9.6)$$

Thus, we have the:

Theorem 9.1. There hold at least one of the following conditions in an *SHR-F*$_n$ admitting an affine motion

$$\mu_l = -\lambda_l\,, \qquad (9.7); \qquad \text{or} \qquad v^j H_j = 0. \qquad (9.8)$$

9.1. Case of Eq. (9.7)

We examine the implications of the necessary consequence of Eq. (9.7) holding in an *SHR-F*$_n$ admitting an affine motion. Such a space is briefly denoted by *ASHR-F*$_n$ [128].The derivatives of the vector-field v^i in the space, given by Eqs. (5.9.1d) and (5.9.5), for Eq. (9.7), thus, reduce to

$$\mathfrak{B}_k\,v^i = -\lambda_k\,v^i, \qquad (9.9a)$$

and

$$\mathfrak{B}_j\,\mathfrak{B}_k\,v^i = (-\mathfrak{B}_j\,\lambda_k + \lambda_j\,\lambda_k)\,v^i. \qquad (9.9b)$$

Next, vanishing of Lie derivative of the Berwald's tensor H^i_{jk} is also seen vide Eqs. (6.8.10) as a necessary condition in such a space. The same, for Eq. (6.2.16b), is written as

$$\pounds H^i_{jk} = v^l\,\mathfrak{B}_l\,H^i_{jk} - H^l_{jk}\,\mathfrak{B}_l\,v^i + H^i_{lk}\,\mathfrak{B}_j\,v^l + H^i_{jl}\,\mathfrak{B}_k\,v^l$$
$$+ (\dot{\partial}_l\,H^i_{jk})\,(\mathfrak{B}_h v^l)\,\dot{x}^h = 0, \qquad (9.10)$$

or, for Eq. (9.9a)

$$v^l\,\mathfrak{B}_l\,H^i_{jk} + \lambda_l\,H^l_{jk}\,v^i - \lambda_j\,H^i_{lk}\,v^l - \lambda_k\,H^i_{jl}\,v^l$$

$$-\lambda_h (\dot{\partial}_l H^i_{jk}) v^l \dot{x}^h = 0,$$

or, for Eq. (2.1) and re-arranging the terms

$$v^l (\mathfrak{B}_l H^i_{jk} + \mathfrak{B}_j H^i_{kl} + \mathfrak{B}_k H^i_{lj}) + v^i \lambda_l H^l_{jk} - v^l H^i_{jkl} \lambda_h \dot{x}^h = 0,$$

or, by Eq. (4.10.13)

$$v^i \lambda_l H^l_{jk} = v^l H^i_{jkl} \lambda_h \dot{x}^h. \tag{9.11}$$

This establishes the:

Theorem 9.2. If an *ASHR-*F_n admits Eq. (9.7), then it also admits Eq. (9.11).

Application of the commutation formula vide Eq. (4.7.14) to the tensor-field $\mathfrak{B}_k v^i$, yields

$$2\,\mathfrak{B}_{[j}\mathfrak{B}_{k]}\mathfrak{B}_h v^i = H^i_{jkl}\,\mathfrak{B}_h v^l - H^l_{jkh}\mathfrak{B}_l v^i - (\dot{\partial}_l \mathfrak{B}_h v^i) H^l_{jk}. \tag{9.12}$$

Interchanging the operators of directional and covariant derivation by Eq. (4.8.3) and noting the independence of the vector v^i on $\dot{x}^l$, the last term on RHS in above equation reduces to $-\, v^m G^i_{lhm} H^l_{jk}$, which on transvection with $\dot{x}^h$ shall be zero for Eq. (4.10.15). Therefore, transvection of Eq. (9.12) by $\dot{x}^h$, for Eqs. (4.6.7) and (4.9.15), yields

$$2\,\mathfrak{B}_{[j}\mathfrak{B}_{k]}\mathfrak{B}_h v^i \dot{x}^h = H^i_{jkl} (\mathfrak{B}_h v^l) \dot{x}^h - H^l_{jk} \mathfrak{B}_l v^i.$$

Putting, from Eq. (9.9a) and using Eq. (9.11), the expression on the RHS of above equation vanishes identically. Thus, we have the:

Theorem 9.3. The covariant derivation processes commute for the vector $\mathfrak{B}_h v^i \dot{x}^h$ in an *ASHR-*F_n :

$$2\,\mathfrak{B}_{[j}\mathfrak{B}_{k]}\mathfrak{B}_h v^i \dot{x}^h = 0. \tag{9.13}$$

For Eqs. (1.1) and (9.9a), it is interesting to note that the tensor field $v^j H^i_{jkh}$ remains constant under the covariant derivation in *ASHR-*F_n:

$$\mathfrak{B}_l (v^j H^i_{jkh}) = (\mathfrak{B}_l v^j) H^i_{jkh} + v^j \mathfrak{B}_l H^i_{jkh}$$

$$= -\lambda_l v^j H^i_{jkh} + v^j \lambda_l H^i_{jkh} = 0. \tag{9.14}$$

Note 9.1. For Eqs. (4.9.15b), (4.9.16) and invariant character of the vector $\dot{x}^h$ under covariant derivation, successive transvections of Eq. (9.14) with $\dot{x}^k, \dot{x}^h$ also make the tensor $v^j H^i_{jk}$ and the vector $v^j H^i_j$ covariant constants:

$$\mathfrak{B}_l\,(v^j H^i_{jk}) = (\mathfrak{B}_l\, v^j) H^i_{jk} + v^j \mathfrak{B}_l\, H^i_{jk} = 0, \tag{9.15a}$$

by Eqs. (1.1) and (9.9a); and

$$\mathfrak{B}_l\,(v^j H^i_j) = (\mathfrak{B}_l\, v^j)\; H^i_j + v^j \mathfrak{B}_l\, H^i_j = 0, \tag{9.15b}$$

Further, contractions of the tensors in above equations w.r.t. the indices i and j, for Eqs. (4.9.20), also yield

$$\mathfrak{B}_l\,(v^j\, H_{kh}) = 0 = \mathfrak{B}_l\,(v^j\, H_k). \tag{9.16}$$

Forming covariant derivative of Eq. (5.9.5), transvecting the resulting equation by $\dot{x}^h$ and putting from Eqs. (5.9.1d), (9.3) together with (9.9a), there results

$$\mathfrak{B}_j\, \mathfrak{B}_k\, \mathfrak{B}_h\, v^i\, \dot{x}^h = 0. \tag{9.17}$$

Putting for $\mathfrak{B}_k\, \mathfrak{B}_h\, v^i$ from Eq. (5.9.5), above equation can be also interpreted as

$$\mathfrak{B}_j\,(v^i \mu_{kh}\, \dot{x}^h) = -(\lambda_j\, v^i)\,(\mu_{kh}\, \dot{x}^h) + v^i \mathfrak{B}_j\,(\mu_{kh}\, \dot{x}^h) = 0,$$

by Eq. (9.9a). For arbitrary vector v^i, it implies

$$\mathfrak{B}_j\,(\mu_{kh}\dot{x}^h) = \lambda_l\,(\mu_{kh}\dot{x}^h). \tag{9.18}$$

Thus, we have the:

Theorem 9.4. The recurrence property of the space *ASHR*-F_n admitting Eq. (9.7) is also shared by the vector $\mu_{kh}\dot{x}^h$.

Forming covariant derivation of Eq. (9.9a) and taking its skew-symmetric part, there also follow the integrability conditions of Eq. (9.9a) by application of Eq. (9.9a) itself:

$$2\mathfrak{B}_{[j}\mathfrak{B}_{k]}\,v^i + 2\,(\mathfrak{B}_{[j}\lambda_{k]})\,v^i = 0, \tag{9.19}$$

or, by the commutation formula vide Eq. (4.7.14) and Eq. (2.5)

$$v^h H^i_{jkh} + v^i A_{jk} = 0. \tag{9.20}$$

§ 10. Integrability conditions of Eq. (9.13)

Let us consider a scalar point function $\sigma\,(x^i)$ satisfying the differential equation

$$\mathfrak{B}_k(\sigma\,\mathfrak{B}_h\,v^i)\,\dot{x}^h = 0, \tag{10.1}$$

and, therefore, also

$$2\mathfrak{B}_{[j}\mathfrak{B}_{k]}\,(\sigma\,\mathfrak{B}_h v^i)\,\dot{x}^h = 0, \quad \text{i.e.} \quad 2\,\sigma\,\mathfrak{B}_{[j}\mathfrak{B}_{k]}\,(\mathfrak{B}_h v^i)\,\dot{x}^h = 0,$$

as $2\mathfrak{B}_{[j}\mathfrak{B}_{k]}\,\sigma = 0$. The function σ being arbitrary (non-zero) the last relation results into Eq. (9.13). Thus, Eq. (10.1) represents an integrability condition for Eq. (9.13).

For Eqs. (9.9) holding in an *ASHR*-F_n, Eq. (10.1) expands to

$$(\lambda_h\,\mathfrak{B}_k\,\sigma + \sigma\,\mathfrak{B}_k\,\lambda_h - \sigma\,\lambda_k\,\lambda_h)\,v^i\,\dot{x}^h = 0.$$

Putting

$$\lambda \equiv \lambda_h\,\dot{x}^h, \tag{10.2}$$

and

$$\sigma_k \equiv (\mathfrak{B}_k\,\sigma)\,/\,\sigma = \mathfrak{B}_k\,(\ln\sigma), \tag{10.3}$$

above equation implies

$$\mathfrak{B}_k\,\lambda + \lambda\,(\sigma_k - \lambda_k) = 0, \tag{10.4}$$

for v^i being arbitrary. This gives rise to the:

Theorem 10.1. There exists a scalar point function σ in an *ASHR*-F_n satisfying Eq. (10.1) and connected to the recurrence vector field λ_k of the space as per Eq. (10.4).

10.1. Particular case

Now we consider a particular case of the integrability condition.) of Eq. (9.13), represented by Eq. (10.1), for $\sigma = 1$:

$$\mathfrak{B}_k\,(\dot{x}^h\,\mathfrak{B}_h\,v^i\,)\;=\;0. \tag{10.5}$$

It may be similarly verified that these conditions also satisfy Eq. (9.13). Proceeding similarly as above, we deduce from Eq. (10.4):

$$\mathfrak{B}_k\,\lambda\;=\lambda\,\lambda_k \tag{10.6}$$

$\Rightarrow$

$$\lambda_k\;=\;(\mathfrak{B}_k\,\lambda)\,/\,\lambda=\mathfrak{B}_k\,(\ln\lambda). \tag{10.7}$$

Further, contraction of Eq. (9.9b) with $\dot{x}^k$, for Eqs. (4.6.7), (10.2) and (10.6), yields back Eq. (10.5):

$$\mathfrak{B}_j\,(\,\dot{x}^k\,\mathfrak{B}_k\,v^i)\;=\;(-\,\mathfrak{B}_j\,\lambda+\lambda_j\,\lambda)\,v^i\;=\;0.$$

Thus, Eq. (10.6) also represents a sufficient condition for Eq. (10.5) as integrability conditions of Eq. (9.13).

Concluding these remarks, we have

Theorem 10.2. The Eq. (10.5) represents integrability conditions of Eq. (9.13) in an *ASHR-*F_n if and only if the recurrence vector λ_k satisfies Eq. (10.7).

Note 10.1. The Eq. (10.5) also expresses the invariance of the vector

$$\dot{x}^h\,\mathfrak{B}_h\;v^i\;=\;-\,(\,\dot{x}^h\,\lambda_h)\,v^i\;=\;\lambda\,v^i\,, \tag{10.8}$$

under the covariant derivation.

§ 11. Some more characteristics of the function σ

Eq. (10.1) involving the function σ also expands to

$$\{(\mathfrak{B}_k\,\sigma)\,\mathfrak{B}_h\;v^i+\sigma\,\mathfrak{B}_k\,\mathfrak{B}_h\;v^i\;\}\;\dot{x}^h\;=0. \tag{11.1}$$

For Eq. (5.9.5) and (9.2), the coefficient of σ in above equation further transforms to

$$\mu_{kh}\;v^i\;\dot{x}^h\;=\;-\,v^j H^i_{jk},$$

which, on transvection with v^k vanishes identically for skew-symmetry of the tensor H^{i}_{jk} in its lower indices. Thus, a transvection of Eq. (11.1) by v^{k} and substitutions from Eqs. (9.9a) and (10.2), yield

$$(v^{k}\mathfrak{B}_k \sigma)\lambda v^{i} = 0 \quad \Rightarrow \quad v^{k}\mathfrak{B}_k \sigma = 0, \tag{11.2}$$

for arbitrary λv^{i}. But, the above expression $v^{k}\mathfrak{B}_k \sigma$ is the Lie derivative of the scalar point function σ, in view of Eq. (6.2.3d). Thus, we have the

Theorem 11.1. The function σ is a Lie-invariant in an *ASHR-*F_n.

Putting from Eq. (10.3), the Eq. (11.1) may be re-written as

$$(\sigma_k \mathfrak{B}_h v^{i} + \mathfrak{B}_k \mathfrak{B}_h v^{i})\dot{x}^{h} = 0. \tag{11.3a}$$

Its covariant derivative w.r.t. x^{j}, for Eq. (9.17), also yields

$$\{(\mathfrak{B}_j \sigma_k)\mathfrak{B}_h v^{i} + \sigma_k(\mathfrak{B}_j \mathfrak{B}_h v^{i})\}\dot{x}^{h} = 0. \tag{11.3b}$$

Elimination of the terms containing second order derivatives of v^{i}, from above equations, there follows another interesting result

$$(\mathfrak{B}_j \sigma_k - \sigma_j \sigma_k)(\dot{x}^{h}\mathfrak{B}_h v^{i}) = 0 \quad \Rightarrow \quad \mathfrak{B}_j \sigma_k = \sigma_j \sigma_k, \tag{11.4}$$

for arbitrary (non-null) vector-field λv^{i} given by Eq. (10.7). Thus, we have:

Theorem 11.2. The vector σ_k defined by Eq. (10.3) shares a recurrence property exhibited by Eq. (11.4) in an *ASHR-*F_n.

The tensor in Eq. (11.4) is symmetric and its further covariant derivation also yields another symmetric tensor

$$\mathfrak{B}_h \mathfrak{B}_j \sigma_k = (\mathfrak{B}_h \sigma_j)\sigma_k + \sigma_j \mathfrak{B}_h \sigma_k = 2\sigma_j \sigma_k \sigma_h, \tag{11.5}$$

by Eq. (11.4) itself. Therefore, the skew-symmetric part of above equation w.r.t. the indices *h* and *j*, for the commutation formula (4.7.14), is

$$2\mathfrak{B}_{[h}\mathfrak{B}_{j]}\sigma_k = -\sigma_l H^{l}_{hjk} - (\dot{\partial}_l \sigma_k) H^{l}_{hj} = 0.$$

The scalar σ being a point function, the vector σ_k is also independent of the directional arguments. Hence, the second term in above equation vanishes and the equation reduces to

$$\sigma_l \, H^l_{h\,j\,k} = 0. \tag{11.6}$$

Theorem 11.3. The relation (11.6) represents integrability conditions for the differential equation (11.4).

§ 12. Implications of Eq. (9.8) in an *SHR-F$_n$*

The directional and successive covariant derivation of Eq. (9.8), for Eq. (4.9.21b), (2.2b), (2.4a) and (9.8) itself, yield

$$\text{(a) } v^j H_{jk} = 0, \quad \text{(b) } (\mathfrak{B}_h v^j)\, H_j = 0, \quad \text{(c) } (\mathfrak{B}_k \mathfrak{B}_h v^j)\, H_j = 0. \tag{12.1}$$

Taking skew-symmetric part of the last equation w.r.t. the indices k and h, and applying the commutation formula in Eq. (4.7.14), there also follows

$$H_j \, H^j_{k\,h\,l} \, v^l = 0. \tag{12.2}$$

It may be noted that either of the Eqs. (12.1) and (12.2) are merely necessary consequences in an *SHR-F$_n$* admitting an affine motion. Either of these equations implies none of Eqs. (6.8.1) and (5.9.1d); hence are not sufficient conditions for an affine motion in *HR-F$_n$*.

Foot-note:

[31] The Theorem 2.2 of [130], indeed holds in a general *HR-F$_n$*.

CHAPTER 9

PROJECTIVE MOTIONS IN RECURRENT FINSLER SPACES

§ 1. Introduction

Projective motions in a general Finsler space F_n are discussed in Chapter 6 with their characterizing equations (6.9.1) - (6.9.3) and integrability conditions represented by Eq. (6.9.6) - (6.9.9). For symmetric and projectively symmetric Finsler spaces these were briefly mentioned in Sub-section 1.2 of Chapter 7. Presently, we discuss the existence of such motions in spaces of recurrent curvature (notably in the sense of Berwald). These were studied by the present author [112], [113] and Sinha [214].

It is seen in the previous chapter (cf. Eqs. (8.2.12) - (8.2.13)) that the projective tensors W^i_{jk} and W^i_j also possess the recurrence properties in an HR-F_n. Hence, the Lie derivation of Eq. (8.2.13), in view of Eq. (6.9.9) holding for a projective motion, yields

$$\pounds\, \mathfrak{B}_l\, W^i_j = (\pounds\, \lambda_l)\, W^i_j. \tag{1.1}$$

Applying the commutation rule vide Eq. (6.6.7) to the tensor W^i_j and putting from Eqs. (5.1.10), (5.2.8), (5.2.17), (6.9.1), (6.9.2) and (6.9.9), there follows

$$W^a_j (p_a\, \delta^i_l + \dot{x}^i p_{al}) - (W^i_l p_j + 2 W^i_j p_l) - (\dot{\partial}_l W^i_j)\, p = (\pounds \lambda_l) W^i_j.$$

On contraction with respect to indices i and l, for Eqs. (5.2.8), (5.2.9), (5.2.12) and homogeneous properties of the function p exhibited by Eqs. (5.1.9) and (5.1.10), it yields

$$(n-2)\, p_l\, W^l_j = (\pounds \lambda_l) W^l_j. \tag{1.2}$$

On the other hand, Lie derivative of Eq. (8.2.12), for Eq. (6.9.8), gives

$$\pounds\, \mathfrak{B}_l\, W^i_{jk} = (\pounds\, \lambda_l)\, W^i_{jk}. \tag{1.3}$$

Interchanging the operators of Lie and covariant derivations as per the commutation rule vide Eq. (6.6.7) applied to the tensor W^i_{jk} and putting from Eqs. (5.2.), (5.2.16), (6.9.1), (6.9.2) and (6.9.8), there follows

$$W^m_{jk}(p_m\,\delta^i_l + \dot{x}^i p_{lm}) + 2W^i_{l[j}p_{k]} - 2W^i_{[j}\,p_{k]l} - pW^i_{jkl}$$

$$= (\pounds\lambda_l + 2p_l)W^i_{jk}. \tag{1.4}$$

Contracting it for indices i and l and simplifying by means of Eqs. (5.1.10), (5.2.19) and (5.2.22), we derive

$$(\pounds\lambda_l - (n-2)\,p_l)W^l_{jk} + 2W^l_{[j}\,p_{k]l} = 0. \tag{1.5}$$

Also, transvection of Eq. (1.4) with $\dot{x}^k\,\dot{x}^l$, for Eqs. (5.1.9), (5.1.10), (5.2.8), (5.2.16) and (5.2.17), yields

$$(\pounds\lambda + 4\,p)W^i_j = \dot{x}^i\,p_m\,W^m_j, \tag{1.6}$$

where we have put

$$\lambda \equiv \lambda_i\,\dot{x}^i \quad\Rightarrow\quad \lambda_i = \dot{\partial}_i\,\lambda, \tag{1.7}$$

for Eq. (8.2.4a). Transvection of Eq. (1.6) by p_i, for Eq. (5.1.9) also gives

$$(\pounds\lambda + 3\,p)\,p_i\;W^i_j = 0. \tag{1.8}$$

§ 2. *PHR-F$_n$*

Definition 2.1. An *HR-F$_n$* admitting a projective motion shall be denoted by *PHR-F$_n$*.

With the introductory formulation in § 1, now we are in position to derive certain necessary conclusions in an *HR-F$_n$* admitting a projective motion.

Theorem 2.1. The recurrence vector λ_i and the vector p_i always satisfy the following relations in a *PHR-F$_n$* :

$$(\pounds\lambda_i)W^i_j = 0 \quad (2.1); \qquad \text{and} \quad p_i\;W^i_j = 0. \tag{2.2}$$

Proof. Assuming if $p_i W^i_j \neq 0$, Eq. (1.8) implies

$$\pounds\lambda + 3\,p = 0\;.$$

The same, on derivation w.r.t. $\dot{x}^i$, for Eqs. (1.7), (5.1.9a) and (6.6.1) yields

$$\pounds\lambda_i + 3\,p_i = 0 \qquad \Rightarrow \qquad \pounds\lambda_i = -3\,p_i;$$

for which, Eq. (1.2) determines

$$(n+1)\,p_l\,W^l_j = 0, \qquad \text{i.e.} \qquad p_l\,W^l_j = 0.$$

This produces a contradiction to the hypothesis - hence establishing Eq. (2.2). Accordingly, Eq. (1.2) establishes Eqs. (2.1). //

For above theorem, Eq. (1.6) yields

$$\pounds\lambda + 4\,p = 0,$$

for arbitrary W^i_j. Differentiating it w.r.t. $\dot{x}^i$, and using the commutation rule vide Eq. (6.6.1), as before, it yields

$$\pounds\lambda_i = -4\,p_i. \tag{2.3}$$

The recurrence vector λ_i being a point function (cf. Theorem 8.2.1), its Lie derivative is found by Eq. (6.2.12a):

$$\pounds\lambda_i = v^j\,(\partial_j\,\lambda_i) + \lambda_j\,(\partial_i\,v^j). \tag{2.4}$$

The generating vector v^i of the projective motion also being a point function, above equation confirms the similar nature of $\pounds\lambda_i$. Hence, the vector p_i is also a point function making its directional derivative given by Eq. (5.1.9b) zero:

$$p_{ij} \equiv \dot{\partial}_i\,p_j = 0. \tag{2.5}$$

Hence, the last term involving the second order derivatives of the function p in Eq. (6.9.1), which characterizes a projective motion in a $HR\text{-}F_n$, vanishes and the equation reduces to

$$\pounds G^i_{jk} = 2\delta^i_{(j}\,p_{k)}. \tag{2.6}$$

Thus, we have the:

Theorem 2.2. If at all there exist projective motions in an $HR\text{-}F_n$ of non-zero scalar curvature, these are of particular form as given by Eq.

(2.6).

Forming the directional derivatives of Eq. (2.6) w.r.t. $\dot{x}^h$ and applying the commutation rule (6.6.12) and Eqs. (3.10.3) and (2.5), one easily obtains the integrability conditions of Eq. (2.6):

$$\pounds\, G^i_{jkh} = 0. \tag{2.7}$$

Further, considering Lie derivative of Eq. (8.2.14), and putting from Eqs. (6.9.6), (6.9.8) and (2.7), there also follows the equation

$$(\pounds\, \mathfrak{B}_l - \pounds\, \lambda_l)\, W^i_{jkh} = 0, \tag{2.8}$$

in a PHR-F_n.

§ 3. Projective motions of special types in *HR*-F_n

Various forms of the vector-field $v^i(x^j)$ generating an infinitesimal transformation vide Eq. (5.4.1) were given by Eqs. (5.9.1). The projective motions of special types: contra, concurrent, special concircular and recurrent ones are studied by the author [113]. It is shown below that an HR-F_n admits neither concurrent nor special concircular projective motions. Also, it is proved that every contra projective motion in non-Riemannian spaces of recurrent curvature are just contra affine motions. However, it is not so in an HR-F_n.

3.1. Parallel vector field

Continuing the discussion of Sub-section of 3.3 of previous chapter, let $\pounds\, \lambda_i$ be a parallel vector field in an HR-F_n ($n > 2$) admitting a projective motion. The covariant derivation of Eq. (2.3) then implies

$$\mathfrak{B}_i\, (\pounds\, \lambda_j) = -4\, \mathfrak{B}_i\, p_j = 0, \tag{3.1}$$

by Eq. (8.3.19). Hence, the vector p_j is also a parallel vector-field.

3.2. Special c.i.t.'s in a recurrent non-Riemannian space A_n*

If the space A_n* considered in Sub-section 6.3 of the previous chapter admits a projective motion characterized by Eqs. (6.3.7) and (6.9.1) together with Eq. (5.9.4), there would hold

$$\pounds G^i_{kh} = \rho_k \delta^i_h + v^j H^i_{jkh} = 2\delta^i_{(k} p_{h)}. \qquad (3.2)$$

Its contraction w.r.t. the indices i and h, for the symmetry of the contracted curvature tensor H_{kh} of A_n* (cf. Eq. (4.10.3)), determines the vector p_k:

$$p_k = n\rho_k / (n+1) . \qquad (3.3)$$

Putting from Eqs. (5.9.14) and (8.6.15), there follow the explicit expressions for p_k:

$$p_k = \{n/(1-n^2)\} H_{kh} v^h = \{n/(n+1)\} \alpha\lambda_k . \qquad (3.4)$$

Thus, we have

Theorem 3.1. The Eqs. (5.9.14), (8.6.15) and (3.4) represent integrability conditions of a special concircular projective motion in an A_n*.

Noting that the discussion of Sub-section 6.3 of previous chapter also holds good here, Eqs. (8.6.18), for Eq. (5.9.1c), assumes the form

$$\pounds p_k = v^j (\mathfrak{B}_j p_k) + p_j (\mathfrak{B}_k v^j) = v^j \mathfrak{B}_j p_k + \rho p_k = 0, \qquad (3.5a)$$

or, for Eqs. (8.3.4) and (8.6.17)

$$v^j \{\mathfrak{B}_j p_k - \lambda_j p_k / 2\} = 0. \qquad (3.5b)$$

Analogously, the Theorem 8.6.2 also holds for the vector-field p_k.

3.3. Special concircular projective motions in *HR*-F_n

Continuing the discussion of Sub-section 9.1 of Chapter 5 in an *HR*-F_n, the Eq. (5.9.16), for Eq. (8.1.1), reduces to

$$(v^h \lambda_l + \rho \delta^h_l)\{H^i_{jkh} - 2\delta^i_{[j} H_{k]h} /(n-1)\} = 0. \qquad (3.6)$$

For Eq. (5.9.15b) and non-null ρ, it yields

$$H^i_{jkh} = 2\delta^i_{[j} H_{k]h} / (n-1). \qquad (3.7)$$

It is seen in Theorem 6.9.3 that a special concircular projective motion with the function ρ as a covariant constant reduces to a contra or con-current projective motion. In continuation with the same, the formula for Lie derivative of the curvature tensor given by Eq. (8.3.3) reduces to the form as in Eq. (8.6.16). The same, on contraction of indices i and j, and transvection with the vector v^k, reduces to

$$£(v^k H_{kh}) = (L + 2\rho)\, v^k H_{kh},$$

where we have also used Eq. (6.2.9a). For Eq. (6.9.13), it further reduces to

$$£\, p_h = (L + 2\rho)\, p_h, \tag{3.8}$$

showing the Lie-recurrence character of the vector p_h. Further transvection of this equation with the vector v^h, for Theorem 6.9.5, yields

$$(L + 2\rho)\, v^h\, p_h = 0$$

$\Rightarrow$

$$\text{(a) either} \quad L + 2\rho = 0, \qquad \text{or} \quad \text{(b)} \qquad v^h\, p_h = 0. \tag{3.9}$$

Thus, for the first possibility, the Lie-invariance of the curvature tensor follows immediately from Eq. (8.6.16).

Summarizing these observations we, thus, have

Theorem 3.2. If an *HR*-F_n admits a special concircular projective motion with the function ρ as a covariant constant, then either its curvature tensor is Lie-recurrent or there holds Eq. (3.9b).

3.4. Contra or concurrent projective motions in *HR*-F_n

In case of contra or concurrent vector field Eqs. (5.9.13) and (5.9.14) reduce to Eqs. (5.9.27) and (5.9.28) respectively. Hence, Eqs. (5.9.15) trivially hold. Covariant derivation of Eq. (5.9.27), for Eqs. (5.9.1a) does not yield any new result for contra field. However, for a concurrent case, similar exercise in *HR*-F_n yields

$$H^{i}_{jkh}\, v^h \lambda_l + c\, H^{i}_{jkl} = 0. \tag{3.10}$$

The same, for Eq. (5.9.27) itself and non-zero c, causes vanishing of the curvature tensor H^{i}_{jkl} - contradicting the very hypothesis of a recurrent space. Hence, we have the

Theorem 3.2. An *HR-*F_n cannot admit a concurrent infinitesimal transformation; and obviously no concurrent projective motions.

However, if it admits a projective motion generated by a contra vector-field satisfying Eqs. (5.9.1a) and (5.9.3), there would follow the identity

$$\pounds\, G^i_{kh} = v^j\, H^i_{jkh} = 2\,\delta^i_{(k}\, p_{h)}, \tag{3.11}$$

from Eqs. (6.3.7) and (2.6). On contraction w.r.t. the indices *i* and *k*, for Eq. (4.9.20), it determines the vector p_k:

$$p_k = -\, v^j H_{jk} \,/\, (n+1).^{33)} \tag{3.12}$$

It may be emphasized that the contracted curvature tensor H_{kh} need not be symmetric in an *HR-*F_n - a contrast to that of a recurrent A_n*. Thus, unlike the special concircular case, where the vector-fields ρ_k and p_k, vanishing together (for Eq. (3.3)), if they do so, p_k is unaffected by vanishing of ρ (for a contra field). Thus, we have the

Theorem 3.3. Every contra projective motion is a contra affine motion in an A_n*; but it need not be so in an *HR-*F_n .

As a further deviation to the case of special concircular projective motion, if an *HR-*F_n admits a contra projective motion, it follows from Eq. (3.12), that the vector-field p_k is recurrent:

$$\mathfrak{B}_j\, p_k = \lambda_j\, p_k. \tag{3.13}$$

It is seen in Sub-section 6.4 of the previous chapter that the curvature tensor (along with its associated tensors and contracted forms) is Lie-recurrent (for $L \neq 0$) for infinitesimal transformations generated by a contra-field and there hold the Eqs. (8.6.19). In case of projective motions, if any, generated by a contra vector-field, there holds Eq. (3.12) and its Lie derivative, for Eqs. (6.2.9a) and (8.6.19), becomes

$$\pounds\, p_k = L\, p_k, \tag{3.14}$$

making the vector p_k also Lie-recurrent.

Alternately, for $L = 0$, the curvature tensor (along with its associated tensors and contracted forms) become Lie-invariant for infinitesimal transformations generated by a contra-field and there hold the Eqs. (8.6.20). Hence, the space admits a curvature collineation. The Lie derivative of the recurrence vector-field λ_k, given by Eq. (8.2.2d), assumes the form

$$£\, \lambda_k \;=\; £\, (\mathfrak{B}_k \ln H). \tag{3.15}$$

Interchanging the operators of Lie and covariant derivations as per the commutation rule vide Eq. (6.6.7), and noting the Lie-invariant nature of the scalar curvature H (cf. Eqs. (8.6.20)), above equation reduces to

$$£\, \lambda_k \;=\; -(\dot{\partial}_i \ln H)\, £\, G^i_k\,,$$

or, for Eqs. (5.9.3) and (6.3.8)

$$£\, \lambda_k \;=\; -(\dot{\partial}_i \ln H)\, v^j\, H^i_{jk}\,, \tag{3.16}$$

holding in an $HR\text{-}F_n$ if admitting a contra projective motion.

Concluding these remarks we have the

Theorem 3.4. An $HR\text{-}F_n$, if admitting a contra projective motion is either Lie-recurrent or admits curvature collineations with its recurrence vector satisfying Eq. (3.16).

3.5. Recurrent projective motion in *HR-F*$_n$

Continuing the discussion of Sub-section 9.2 in Chapter 6, we now explore recurrent projective motions in a recurrent Finsler space where-in hold the additional conditions expressed by Eqs. (8.1.1), (8.2.1) - (8.2.2) simultaneously. Covariant derivations of Eqs. (6.9.24) and (6.9.25), for Eqs. (5.9.1d) and (8.2.2) yield

$$\mathfrak{B}_k\, p \;=\; \zeta_k\, p \quad (3.17);\ \text{and} \quad \mathfrak{B}_k\, p_h \;=\; \zeta_k\, p_h, \tag{3.18}$$

where

$$\zeta_k \equiv \mu_k + \lambda_k\,. \tag{3.19}$$

Eqs. (3.17) and (3.18) express the recurrent properties of the function p and its derivative p_h. Further, the directional derivative of Eq. (3.17), for commutation formula (4.8.3) used for the scalar function p and (5.1.9a), yields

$$\mathfrak{B}_k\, p_h \;=\; (\dot{\partial}_h\, \zeta_k)\, p + \zeta_k\, p_h,$$

or, by Eqs. (8.2.4a), (3.18)

$$(\dot{\partial}_h \mu_k)\, p = 0 \quad \Rightarrow \quad \dot{\partial}_h \mu_k = 0, \tag{3.20}$$

for arbitrary p; which is essentially non-zero for the existence of a projective motion.

Hence, Eq. (5.10.2) reduces to the form of Eq. (5.9.12) holding for a special c.i.t. Further, Eq. (3.20) reduces the covariant derivative of μ_k, obtainable by Eq. (4.6.6), as

$$\mathfrak{B}_j \mu_k = \partial_j \mu_k - \mu_i G^i_{jk} \quad \Rightarrow \quad 2\mathfrak{B}_{[j} \mu_{k]} = 2\partial_{[j} \mu_{k]}, \tag{3.21}$$

for symmetry of G^i_{jk} in its lower indices. Therefore, Eq. (5.10.1) reduces to

$$H^i_{jkh} v^h = 2 v^i \partial_{[j} \mu_{k]}. \tag{3.22}$$

Above remarks are summarized in the following theorem:

Theorem 3.5. If an *HR-*F_n of non-zero scalar curvature admits a recurrent projective motion, there would hold the following necessary consequences:

(i) both the scalar function p and the vector-field p_k are recurrent;

(ii) like the recurrence vector λ_i the vector field μ_j becomes a point function; and

(iii) the Eqs. (5.10.1) and (3.22) represent alternate forms of integrability conditions.

For Eq. (3.20), the Bianchi identity represented by Eq. (8.6.23) in an *HR-*F_n admitting a recurrent projective motion reduces to

$$\lambda_{[j} H^i_{kh]m} v^m = 0, \tag{3.23a}$$

or, for Eq. (3.22) and arbitrary v^i

$$\lambda_{[j} \partial_k \mu_{h]} = 0. \tag{3.23b}$$

In order to derive an explicit expression for the vector μ_j, we establish a lemma:

Lemma 3.1. Lie derivative of the scalar curvature of an $HR\text{-}F_n$ admitting a recurrent projective motion is given by

$$\pounds H + p\,\zeta = 0, \tag{3.24}$$

where

$$\zeta \equiv \dot{x}^k \zeta_k = \dot{x}^k (\mu_k + \lambda_k) = \mu + \lambda, \tag{3.25}$$

by Eqs. (5.10.4) and (8.10.2).

Proof. Lie derivative of the curvature tensor is given by Eq. (6.8.9). Transvecting it by $\dot{x}^h$, and using Eqs. (4.6.2a), (4.6.7c), (4.9.15b), (4.10.15) and (6.2.9b), there follows

$$\pounds H^i_{jk} = \mathfrak{B}_j (\pounds G^i_k) - \mathfrak{B}_k (\pounds G^i_j) = 2\mathfrak{B}_{[j} (p\,\delta^i_{k]} + \dot{x}^i p_{k]}), \tag{3.26}$$

by Eq. (6.9.2). Its further transvection by $\dot{x}^k$, for Eqs. (4.9.16) and (5.1.10a), yields

$$\pounds H^i_j = 2\,\dot{x}^i \mathfrak{B}_j p - \dot{x}^k \mathfrak{B}_k (p\,\delta^i_j + \dot{x}^i p_j). \tag{3.27}$$

Contracting it w.r.t. the indices i and j, for Eqs. (4.9.20), (5.1.10a), (3.17) and (3.25), we finally get the desired result. //

But, Lie derivative of the scalar curvature H can also be evaluated by formula (6.2.3d):

$$\pounds H = v^i (\mathfrak{B}_i H) + (\dot{\partial}_i H)(\mathfrak{B}_k v^i)\,\dot{x}^k = L\,H + \mu\, v^i\, \dot{\partial}_i H, \tag{3.28}$$

by Eqs. (5.9.1d), (5.10.4), (8.2.2c) and (8.3.4). Comparing Eqs. (3.24) and (3.28) and putting from Eq. (3.25), we derive

$$\mu\,(p + v^i\,\dot{\partial}_i H) + (L\,H + \lambda\,p) = 0. \tag{3.29}$$

For Eq. (6.9.24), the coefficient of μ in above equation is

$$v^i \{H_i + (1-n)\dot{\partial}_i H\}/(1-n) = -v^i\,\dot{x}^k H_{ki}/(1-n),$$

by Eq. (4.9.22). This being non-zero, in general, Eq. (3.29) determines the function

$$\mu = (1-n)(L\,H + \lambda\,p)/\,v^i\,\dot{x}^k H_{ki}. \tag{3.30}$$

Thus, we have the

Theorem 3.6. Eqs. (6.9.24) and (3.30) determine the complete integrability conditions for a recurrent projective motion in an *HR*-F_n of non-zero scalar curvature.

§ 4. Some more implications of a recurrent projective motion in an *HR*-F_n

It is seen above that the curvature tensor of an *HR*-F_n space admitting a recurrent projective motion necessarily satisfies Eq. (3.22). Contracting this equation w.r.t. the indices i and j, and using Eqs. (4.9.20a), we get

$$H_{kh}\, v^h = 2v^j\, \partial_{[j}\, \mu_{k]}. \tag{4.1}$$

Also, a transvection of Eq. (3.22) with v^j, and use of Eq. (4.1), yields

$$H^i_{jkh}\, v^j\, v^h = H_{kh}\, v^i\, v^h. \tag{4.2}$$

On the other hand, a transvection of Eq. (4.1) with v^k, gives

$$H_{kh}\, v^k\, v^h = 2\, v^j\, v^k\, \partial_{[j}\, \mu_{k]} = 0, \tag{4.3}$$

for skew-symmetric properties of the RHS. This makes the tensor H_{kh} also skew-symmetric in its indices:

$$H_{kh} = -\, H_{hk}. \tag{4.4}$$

Consequently, identities vide Eqs. (4.9.21), (4.10.3) etc. are also written as

$$H_{kh}\, \dot{x}^k = -\, H_h, \qquad H_{kh} = -\dot{\partial}_k\, H_h, \qquad H^r_{jkr} = 2\, H_{kj}. \tag{4.5}$$

Hence, the projective tensors given by Eqs. (5.2.13c) and (5.2.15) reduce to

$$W^i_{jk} = H^i_{jk} + 2\,(\dot{x}^i\, H_{jk} + H_{[j}\, \delta^i_{k]})\,/\,(n+1), \tag{4.6}$$

and

$$W^i_{jkh} = H^i_{jkh} + 2\,(\dot{x}^i\, \dot{\partial}_h\, H_{jk} + H_{jk}\, \delta^i_h - H_{h[j}\, \delta^i_{k]})\,/\,(n+1), \tag{4.7}$$

by Eqs. (4.5). Transvecting Eq. (4.7) with $v^j\, v^h$ and simplifying by means of Eqs. (4.2) - (4.4) and the identity resulting from Eqs. (4.9.21b) and (4.3):

$$v^j\, v^h\, \dot{\partial}_h\, H_{jk} = v^j\, v^h\, \dot{\partial}_k\, H_{jh} = \dot{\partial}_k\, (v^j\, v^h H_{jh}) = 0,$$

we derive

$$W^{i}_{jkh}\, v^{j}\, v^{h} = H^{i}_{jkh} v^{j}\, v^{h} + (2H_{jk}\, v^{i}\, v^{j} + H_{hk}\, v^{i}\, v^{h}) / (n+1)$$

$$= \{H_{kh} + 3H_{hk} / (n+1)\}\, v^{i}\, v^{h}, \quad \text{by Eqs. (4.2)}$$

$$= \{(n-2)/(n+1)\}\, H_{kh}\, v^{i}\, v^{h}, \quad \text{by Eqs. (4.4)} \qquad (4.8a)$$

$$= 2\{(n-2)/(n+1)\}\, v^{i}\, v^{j}\, \partial_{[j}\, \mu_{k]}, \qquad (4.8b)$$

by Eqs. (4.1). As exhibited by Eq. (6.9.6), the Lie derivative of the projective curvature tensor is a necessary consequence of a projective motion in an F_n. Thus, it follows from above equation that the vector-field $H_{kh}\, v^{h}$ is Lie-invariant in an HR-F_n admitting a recurrent projective motion:

$$\pounds\, (H_{kh}\, v^{h}) = 0. \qquad (4.9)$$

Thus, we have the:

Theorem 4.1. Contracted curvature tensor H_{kh} of an HR-F_n admitting a recurrent projective motion is skew-symmetric; and its transvected part with v^{h} is a Lie-invariant.

4.1. Gradient vector-field μ_k

If, in addition, the vector-field μ_k satisfies

$$2\partial_{[j}\, \mu_{k]} = 0, \quad \text{i.e.} \qquad \partial_j\, \mu_k = \partial_k\, \mu_j, \qquad (4.10)$$

the differential equation $\mu_j\, dx^{j} = 0$ becomes exact implying the gradient property of the vector-field μ_j:

$$\mu_j \equiv \partial_j\, \omega, \qquad (4.11)$$

for some non-null scalar function ω. Hence, the Eqs. (3.22) and (4.1) reduce to Eqs. (5.9.27) and (5.9.28) respectively considered in Subsection 9.3 of Chapter 5 for special c.i.t. Besides, Eq. (5.9.12) also holds good in an HR-F_n admitting a recurrent projective motion in view of Eqs. (5.10.2) and (3.20). Thus, it is to be noted that under the hypothesis of Eq. (4.10), a recurrent projective motion characterized by Eqs. (5.9.1d) and (6.9.1) satisfies all the three Eqs. (5.9.12), (5.9.27) and (5.9.28), which characterize a contra projective motion in an HR-F_n.

CHAPTER 10

GROUPS OF TRANSFORMATIONS IN FINSLERIAN SPACES (Revisited)

§ 1. Introduction

Infinitesimal transformations defining motions, affine motions, projective motions, conformal transformations and curvature collineations in various types of Finslerian spaces have been discussed in the previous chapters. This study is revisited here with some additional remarks.

§ 2. Motions

Infinitesimal transformations vide Eq. (5.4.1) generated by vector-field $v^i(x^j)$ were considered. Necessary and sufficient conditions for a motion were derived in § 7 of Chapter 6. The author with R.S. Mishra [95], [138] also considered *generalized infinitesimal transformations* generated by a vector field v^i depending upon the line-element $(x^i, \dot{x}^i)$ and motion generated by such (generalized) transformation has been studied in a Finsler space [95], [105]. Matsumoto [82], [84] and Yasuda [239] also considerd a generalized infinitesimal transformation taking the vector field v^i as a function of line-element (x^i, y^i) and applied it to different kinds of Finslerian structures.

§ 3. Affine motions

3.1. Infinitesimal transformations preserving the parallelism of a pair of vectors defined an affine motion (§ 8 of Chapter 6) and necessary and sufficient condition for the same were derived vide Eq. (6.8.1). Matsumoto [70], [83] and others [189], [234] also studied affine motions generated by infinitesimal transformation of restricted type, i.e. vide Eq. (5.4.1) in a general Finsler space.

3.2. Taking vanishing covariant derivative of Berwald's curvature tensor a *symmetric Finsler space* in the sense of E. Cartan, denoted as $S\text{-}F_n$, was defined by the author [108] in Chapter 7. Affine motions generated by infinitesimal transformations of restricted type vide Eq. (5.4.1) were considered there and special attention focussed for the recurrent vector-field v^i satisfying Eq. (5.9.1d). Affine motions generated by contra, concurrent and special concircular vector-fields satisfying Eqs. (5.9.1.a) - (5.9.1c) are studied by the author et al. [96] in symmetric as well as a general Finsler space.

3.3. Finslerian spaces of recurrent curvature were discussed in Chapter 8, where the covariant derivative of the curvature tensor is proportional to the tensor itself. These were defined by Eq. (8.1.1) and denoted by HR-F_n. Infinitesimal transformations vide Eq. (5.4.1) generated by the vector-field of different types discussed in § 9 of Chapter 5 were considered and affine motions were considered in detail in § 8 of Chapter 6.

3.4. Bi-recurrent Finsler spaces, for which the curvature tensor possesses second order recurrent property:

$$\mathfrak{B}_{l\,m}\, H^{\;i}_{j\,k\,h} = a_{l\,m}\, H^{\;i}_{j\,k\,h}, \tag{3.1}$$

for some non-null covariant tensor field $a_{l\,m}$, were defined and studied by Moór [148], Sinha and Singh [210], [211] and the present author with Meher [133]. An n-dimensional Finsler space with Eq. (3.1) is denoted by HR^2-F_n [133]. Affine motions in such spaces have been studied by the present author [111] and the corresponding space is denoted by AHR^2-F_n [111]. Particular cases when the vector field v^i generating an affine motion is of concurrent or special concircular form satisfying Eqs. (5.9.1b) and (5.9.1c) have been considered. Following are some of the results established therein [111], Theorems 3.2, 4.1.

Theorem 3.1. A bi-recurrent Finsler space admitting a special concircular affine motion [32)] reduces to an HR-F_n .

Theorem 3.2. There exist neither contra nor concurrent affine motions in an HR^2-F_n.

3.5. Analogous to Soós [217] concept of projectively symmetric Riemannian space the present author defined projectively symmetric Finslerian space by taking vanishing covariant derivatives of projective curvature tensor. An n-dimensional Finsler space whose projective curvature tensor possesses vanishing covariant derivative satisfying Eq. (7.5.1) is denoted as PS-F_n [109].

Special concircular affine motions characterized by Eqs. (6.8.1) and (5.9.1c) in a PS-F_n were discussed by the author with Meher [134] in Sub-section 1.1 and § 10 of Chapter 7.

§ 4. Projective motions

4.1. Infinitesimal transformations of a Finsler space F_n over another Finsler space $\overline{F}_n$ mapping geodesics of F_n over geodesics of $\overline{F}_n$ define projective motions. A necessary and sufficient condition for a projective motion has been derived by the present author with Meher [131] in Chapter 6 vide Eq. (6.9.1).

Analogous to *PS-*F_n (introduced by the present author [109]) Pande and Pandey [169] considered vanishing projective covariant derivative of projective entities Q^{i}_{jkh} introduced by the author [98] and defined a projectively symmetric Finsler space so-called "*special projectively symmetric Finsler space*". Projective motions have been studied by them in such a space.

4.2. Meher [88] and Pandey [180] studied projective motions in a symmetric Finsler space. They derive certain conditions for the existence of projective motions in such spaces.

4.3. Projective motions in Finsler spaces of recurrent curvature (in the sense of Berwald), studied by the author [112], [113] and Sinha [214], are dealt with in Chapter 9.

Projective motions generated by vector-field of special types discussed in § 9 of Chapter 5 are included.

4.4. Pandey [180] also studied projective motions in a projectively symmetric Finsler space and proved the following result:

Theorem 4.1. A non-trivial projectively symmetric Finsler space cannot admit any proper projective motions, instead, the projective motion reduces to an affine motion in such a space.

4.5. Normal projective connection parameters

$$\overset{(n)}{\Pi}{}^{i}_{jk} \equiv G^{i}_{jk} - \dot{x}^{i} G^{a}_{jka} / (n+1), \tag{4.1}$$

have been introduced by Yano [234] and the space F_n equipped with these has been called a "*normal projective Finsler space*". The present author with Nawal-Kishore and Pandey [141] defined a symmetric normal projective Finsler space (denoted by $SNP\text{-}F_n$) by taking vanishing covariant derivative of the normal projective curvature tensor N^{i}_{jkh}:

$$\nabla_l N^{i}_{jkh} = 0, \tag{4.2}$$

where ∇_l denotes the covariant differential operator with respect to the normal projective connection parameters defined by Eq. (4.1) and N^{i}_{jkh} the corresponding curvature tensor. Projective motions have been studied by us [141] in $NP\text{-}F_n$ and $SNP\text{-}F_n$, and the following results are established [Theorem 2.1, Eqs. (4.6) and (4.7)]

Theorem 4.2. The normal projective curvature tensor is Lie-invariant in an $NP\text{-}F_n$ admitting a projective motion characterized by an additional condition $\nabla_k\, p = 0$.

Theorem 4.3. $\pounds\, W^{i}_{jk} = 0$ and $p_{lh}\, W^{h}_{jk} = 0$ determine the integrability conditions for the existence of a projective motion in an $SNP\text{-}F_n$.

4.6. The present author with Meher [127] and Singh [203] defined a *normal projective Finsler space of recurrent curvature* if its normal projective curvature tensor N^{i}_{jkh} satisfies

$$\nabla_l N^{i}_{jkh} = \lambda_l\, N^{i}_{jkh}, \tag{4.3}$$

for some non-null covariant vector field $\lambda_l\,(x, \dot{x})$. Such an n-dimensional Finsler space is denoted by $RNP\text{-}F_n$ [127]. Existence of projective motions in an $RNP\text{-}F_n$ has been discussed independently by above authors and the following results have been concluded [127], Theorems 3.1, 4.3:

Theorem 4.4. An $RNP\text{-}F_n$, $(n > 2)$, admits a projective motion characterized by

$$p_k = (n-2)^{-1} \pounds\, \lambda_k . \tag{4.4}$$

Theorem 4.5. Lie-invariance of the recurrence vector λ_i is a necessary and sufficient condition for a projective motion to be an affine motion in an $RNP\text{-}F_n$, $(n > 2)$.

§ 5. Curvature collineations

5.1. Curvature collineations defined by Lie invariance of the curvature tensor were considered in § 10 of Chapter 6.

Theorem 5.1. There exist no no-trivial *H-CC* projective motions in a symmetric Finsler space. Indeed, it reduces to an *H-CC* affine motion.

Theorem 4.3 of [140] deals with the integrability conditions of an *H-CC* projective motion in a $PS\text{-}F_n$, $(n > 2)$, and $WR\text{-}F_n$.

Pandey [172] considers curvature collineations of special concircular form [33)], characterized by Eq. (5.9.1c), in a bi-recurrent Finsler space, and establishes the following result:

Theorem 5.2. A bi-recurrent Finsler space admitting curvature collineations generated by a special concircular vector field v^i reduces to a recurrent Finsler space.

5.2. Projective curvature collineations (denoted by *P-CC* and characterized by Eq. (6.8.12) in a Finsler space have been studied by the present author with Nawal-Kishore [140].

§ 6. Summary of groups of transformations in different Finslerian spaces

Note 6.1. Numbers in the table (excluding the 1st row) refer to the references in the bibliography.

Sr. Nos.	**1**	**2**	**3**	**4**	**5**	**6**	**7**	**8**	**9**	**10**	**11**	**12**
Types of spaces → ↓ Trans-formations	F_n	$P\text{-}F_n$	$S\text{-}F_n$	$HR\text{-}F_n$	$HR\text{-}F_n$ (revi-sed)	$HR^2\text{-}F_n$	$PS\text{-}F_n$	$NP\text{-}F_n$	$SNP\text{-}F_n$	$RNP\text{-}F_n$	$WR\text{-}F_n$	$WR^2\text{-}F_n$
Defining charact-eristics		131	108	97, 110, 130, 147, 196, 213	89, 112, 113	133, 148, 210, 211	109	234	141	127, 203	144	168
Motion	189, 234											
Generalized motion	95, 105											
Affine motion	70, 189, 234		96	64, 130, 213	112, 176	111						
Contra	96		96			111						

affine motion												
Concurrent affine motion			96			111						
Special concircular affine motion			96			111	134					
Recurrent affine motion			108	128, 174	112							
Projective motion		131, 169	88, 180	214	112		180	141	141	127, 203		
Contra projective motion	170				113							
Concurrent projective motion					113							
Special concircular projective motion	132			132	113							
Recurrent projective motion					113							
Conformal transformation	48-53, 99, 102, 106, 126, 171, 173, 182, 189							182				
Curvature collineation	140, 167, 183, 207, 205		96, 140				140				140	
Special concircular curvature collineation						172						
Projective curvature collineation	140		169									
Concurrent inf. transfor.	85											
Concircular inf. transf.	2, 52, 120				135, 175							
Special concircular affine motion			96			111	134					
Recurrent affine motion			108	128, 174	112							
Projective motion		131, 169	88, 180	214	112		180	141	141	127, 203		
Contra projective motion	170				113							
Concurrent projective motion					113							
Special concircular projective motion	132			132	113							
Recurrent projective motion					113							
Conformal transfor-	48-53,							182				

mation	99, 102, 106, 126, 171, 173, 182, 189											
Curvature collineation	140, 167, 183, 208, 205		96, 140				140				140	
Special concircular curvature collineation						172						
Projective curvature collineation	140		169									
Concurrent inf. transfor.	85											
Concircular inf. transf.	2, 52, 120				135, 175							

Foot-notes:

[32] Cf. Section 6 of this chapter.

[33] Called *CA*-collineations by him.

CHAPTER 11

ON PROJECTIVELY FLAT FINSLERIAN SPACES

§ 1. Introduction

Projectively flat Finslerian spaces were, for the first time, introduced by Douglas [39]. Later, Berwald [21] studied the two-dimensional Finslerian spaces of vanishing projective curvature. Okumura [161] exhibited certain properties of projectively flat non-Riemannian spaces admitting concircular and torse-forming vector fields in association with symmetric and recurrent properties of their curvature tensors. Meher [89] considered the concept of projective flatness in a Finsler space whose Berwald's curvature tensor possesses a recurrent property and he derived certain relations connecting the curvature tensor and the recurrence vector. Pandey [178] derived a necessary and sufficient condition for the projective flatness of a Finsler space in terms of its isotropic property. In the present chapter, we study the projective flat Finslerian spaces F_n , $n > 2$, in association with their sectional curvature, symmetric and recurrent character of their curvature, and normal projective curvature tensor.

§ 2. Projectively flat F_n

The concept of projective flatness in Finsler geometry is given in terms of its vanishing projective curvature tensor:

$$W^i_{jkh} = 0 \quad \Rightarrow \quad W^i_{jk} = 0, \quad \text{and} \quad W^i_j = 0, \qquad (2.1)$$

on successive transvections with $\dot{x}^h, \dot{x}^k$ and using Eqs. (5.2.16) and (5.2.17). It follows from Eqs. (5.2.15) and (5.2.13c) that the Berwald's tensors accordingly satisfy

$$H^i_{jkh} = (H^r_{jkr}\,\delta^i_h + \dot{x}^i \dot{\partial}_h H^r_{jkr})/(n+1)$$

$$-2(n\,\dot{\partial}_h H_{[j} + H_{h[j} + \dot{x}^r\,\dot{\partial}_h H_{r\,[j})\,\delta^i_{k]}/(n^2-1), \qquad (2.2)$$

together with

$$H^i_{jk} = (\dot{x}^i H^r_{jkr})/(n+1) - 2\{n H_{[j} + \dot{x}^r H_{r\,[j}\}\delta^i_{k]}/(n^2-1)\,, \qquad (2.3)$$

in a projectively flat F_n. Substitution

$$B_{kh} = (n\,H_{kh} + H_{hk})/(1 - n^2), \tag{2.4}$$

for Eq. (4.9.21b), reduces Eq. (2.2) to the form

$$H^i_{jkh} = (H^r_{jkr}\,\delta^i_h + \dot{x}^i\dot{\partial}_h\,H^r_{jkr})/(n+1)$$
$$-2\,\delta^i_{[j}\,B_{k]h} - 2\left\{\dot{x}^a\dot{\partial}_h H_{a[j}\right\}\delta^i_{k]}\,/(n^2-1). \tag{2.5}$$

In F_n with symmetric H_{kh}, from Eqs. (4.10.3) and (2.4), there follow

$$H^i_{jki} = 0 \quad (2.6a); \qquad B_{kh} = H_{kh}\,/\,(1-n), \quad (2.6b)$$

respectively, whereas the last term in the right side of Eq. (2.5) vanishes for Eq. (4.9.21b) and the homogeneous properties of the tensor H_{jh} :

$$\dot{x}^a\,\dot{\partial}_h H_{aj} = \dot{x}^a\,\dot{\partial}_h H_{ja} = \dot{x}^a\,\dot{\partial}_h\dot{\partial}_a H_j = \dot{x}^a\,\dot{\partial}_a H_{jh} = 0. \tag{2.7}$$

Consequently, Eq. (2.5) reduces to

$$H^i_{jkh} = 2\,\delta^i_{[j}\,H_{k]h}/\,(n-1), \tag{2.8}$$

in such a space, which is precisely the Eq. (3.5) of [135]. As seen in [135], p.290, the Eq. (2.8) also implies the symmetry of H_{kh} as well as the projective flatness of the space F_n, $n > 2$. Thus, we have the

Theorem 2.1. Equation (2.8) represents a necessary and sufficient condition for a projectively flat F_n, $n > 2$, with a symmetric contracted curvature tensor H_{kh}.

A Finsler space F_n, $n > 2$, of constant sectional curvature R is characterized by necessary and sufficient conditions vide Eqs. (4.11.7) and (4.11.9) ([189], Section 7). Eq. (4.11.7), on transvection by $\dot{x}^k$, for Eqs. (3.10.2), (4.2.3), (4.9.16) and (4.11.8), also yields

$$H^i_j = F^2 R\,(\delta^i_j - l^i l_j). \tag{2.9}$$

Further, its contraction w.r.t. the indices i and j, for Eq. (4.9.20c), implies Eq. (4.11.5). Using Eqs. (4.11.9) and (4.11.10), it is seen in [120], Section 5, that such a space necessarily admits Eq. (2.8). In order to check the sufficiency of the condition vide Eq. (2.8) to make the space of constant sectional curvature we proceed as follows. Transvecting Eq. (2.8) by $\dot{x}^h$, and applying Eqs. (4.9.15b) and (4.9.21a) we have

$$H^i_{jk} = 2\,\delta^i_{[j}\, H_{k]} / (n-1). \tag{2.10a}$$

Its further transvection by l_i, for Eq. (4.9.23), yields

$$l_j\, H_k = l_k\, H_j, \tag{2.11}$$

which, on transvection, by $l^j = \dot{x}^j / F$, for Eqs. (4.2.5) and (4.9.21c), gives

$$H_k = (n-1)\,(H/F)\, l_k. \tag{2.12}$$

Accordingly, Eq. (2.10a) reduces to

$$H^i_{jk} = 2\,(H/F)\;\delta^i_{[j} l_{k]}, \tag{2.10b}$$

which does not coincide with Eq. (4.11.7) unless there holds Eq. (4.11.5). Assuming equivalence of H/F^2 and the sectional curvature R of an F_n, $n > 2$, (which is so when R is constant [189], p. 147), Pandey [175] hastily concludes Eq. (4.11.7) from Eq. (2.10b) and establishes the sufficiency of the condition vide Eq. (2.8) for a space of constant sectional curvature. A substitution for $(H/F)l_k$ from Eq. (2.12) in Eq. (2.10b) concludes Eq. (2.10a); which, on differentiation with respect to $\dot{x}^h$, yields Eq. (2.8) back, for Eqs. (4.9.15a) and (4.9.21b). Thus, we have established the

Theorem 2.2. The equation (2.8) represents only a necessary condition for an F_n, $n > 2$, of constant sectional curvature. However, Eq. (2.10b) becomes a necessary and sufficient condition for Eq. (2.8).

From Theorems 2.1 and 2.2 we, therefore, conclude that Eq. (2.10b) becomes a necessary and sufficient condition for a projectively flat F_n, $n > 2$, with a symmetric tensor H_{kh}. Thus, we have the

Corollary 2.1. A projectively flat F_n, $n > 2$, admits Eq. (2.10b) iff its contracted curvature tensor H_{kh} is symmetric.

In the following we characterize a projectively flat Finsler space by an explicit form of its connection parameters. Let p_k and p_{kh} be two arbitrary sets of functions defining the connection parameters:

$$G^i_{kh} \equiv 2\,\delta^i_{(k}p_{h)} + \dot{x}^i p_{kh}, \tag{2.13}$$

which satisfy their usual laws of coordinate transformation. As G^i_{kh} and the first term on the right side of Eq. (2.13) are symmetric in their lower indices the functions p_{kh} are also of the same character. Furthermore, the homogeneous properties of G^i_{kh} require that the functions p_k and p_{kh} should be positively homogeneous of degree zero and -1 respectively in their directional arguments, if any. In the following, we establish that the two sets of functions are not independent to each other. Indeed, the latter ones are derivable from the former ones. It is also proved that the functions p_k are directional derivatives of some scalar function, say p.

To do this we, first, transvect Eq. (2.13) by $\dot{x}^h$ and use Eq. (4.6.2a):

$$G^i_k = \delta^i_k p_h \dot{x}^h + \dot{x}^i p_k + \dot{x}^i p_{kh}\dot{x}^h. \tag{2.14}$$

Its directional differentiation w.r.t. $\dot{x}^h$, in view of Eqs. (4.6.1), (2.13) and symmetry of p_{kh}, yields

$$\left(\delta^i_k \dot{\partial}_h p_l + \delta^i_h p_{kl}\right)\dot{x}^l + \dot{x}^i\left(\dot{\partial}_h p_k + \dot{x}^l \dot{\partial}_h p_{kl}\right) = 0. \tag{2.15}$$

When contracted for i and h, and homogeneous properties of the functions p_k and p_{kh} used, above equation yields

$$\{\dot{\partial}_k p_l + (n-1)p_{kl}\}\,\dot{x}^l = 0. \tag{2.16}$$

Transvecting it further by $\dot{x}^k$, and using homogeneous properties of p_l again, we obtain

$$p_{kl}\dot{x}^k\dot{x}^l = 0,$$

for $n \neq 1$. Accordingly, a transvection of Eq. (2.13) with $\dot{x}^k\dot{x}^h$, for Eqs. (4.6.2a) and (4.6.3) determines

$$G^i = \dot{x}^i p_h \dot{x}^h. \tag{2.17}$$

Again, a directional differentiation of Eq. (2.17), for Eqs. (4.6.1) and (2.14), implies

$$p_{kl}\dot{x}^l = (\dot{\partial}_k\, p_l)\dot{x}^l, \tag{2.18}$$

for arbitrary $\dot{x}^i$'s. For the consistency of Eqs. (2.16) and (2.18) we, must, therefore have

$$p_{kl}\,\dot{x}^l = (\dot{\partial}_k\, p_l)\,\dot{x}^l = 0, \tag{2.19}$$

together with their directional derivatives:

$$(\dot{\partial}_h\ \ p_{kl})\dot{x}^l + p_{kh} = 0. \tag{2.20}$$

An application of the last two relations in Eq. (2.15), thus, determines

$$p_{kh} = \dot{\partial}_h p_k = \dot{\partial}_k p_h, \tag{2.21}$$

for arbitrary $\dot{x}^i$'s and symmetry of p_{kh}. This establishes the exact property of the differential equation $p_k\ d\dot{x}^k = 0$ having a solution

$$p_k = \dot{\partial}_k p, \tag{2.22}$$

for some scalar function p which should be positively homogeneous of degree 1 in its directional arguments. From Eq. (2.22), we also derive

$$p_k\ \dot{x}^k = p. \tag{2.23}$$

Consequently, Eqs. (2.14) and (2.17) reduce to

$$G^i_k = p\,\delta^i_k + \dot{x}^i p_k \quad \text{(2.24a)}; \qquad G^i = p\,\dot{x}^i, \tag{2.24b}$$

where Eq. (2.19) is also used. Also, a directional derivation of Eq. (2.13), in view of Eq. (2.21), determines the tensor

$$G^i_{jkh} \equiv \delta^i_k p_{hj} + \delta^i_h p_{jk} + \delta^i_j p_{kh} + \dot{x}^i p_{jkh}, \tag{2.25}$$

where

$$p_{jkh} = \dot{\partial}_j p_{kh}, \tag{2.26}$$

satisfies

$$\dot{x}^j p_{jkh} = -p_{kh}. \tag{2.27}$$

With such characteristics of the functions p_k and p_{kh} we now look for the corresponding nature of the curvature tensors and projective curvature tensors of the space. Putting from Eqs. (2.13), (2.24a) and (2.25) in Eq. (4.7.15), and using homogeneous properties of the functions p_k and p_{kh} we derive after some simplification

$$H^i_{jkh} = 2\{\partial_{[j}\, p_{k]}\delta^i_h - \delta^i_{[j}\left(\partial_{k]} - p_{k]}\right)p_h\} + 2\{\dot{x}^i\partial_{[j} + p\delta^i_{[j}\}\, p_{k]h}. \tag{2.28}$$

Its contraction w.r.t. the indices *i*, *j*, and *i*, *h*, for Eqs. (4.9.20a) and (4.10.3) also determines

$$H_{kh} = (\partial_h\, p_k - n\partial_k p_h) + (n-1)(\, p_k p_h + pp_{kh}) + \dot{x}^i\partial_i\, p_{kh}. \tag{2.29}$$

together with

$$H^a_{jka} = 2\,(n+1)\;\partial_{[j}\, p_{k]}\,. \tag{2.30}$$

Accordingly, Eq. (2.4) assumes the form

$$B_{kh} = \partial_k p_h - (\, p_k p_h + pp_{kh}) + (\dot{x}^i\partial_i\, p_{kh})\,/\,(1-n). \tag{2.31}$$

Also, directional derivations of Eqs. (2.29) and (2.30), for Eqs. (2.19), (2.21) - (2.23) and (2.27), yield

$$\dot{x}^a\dot{\partial}_h H_{aj} = -(n+1)\;\dot{x}^a\partial_a\, p_{hj}, \tag{2.32}$$

and

$$\dot{\partial}_h H^a_{jka} = 2\,(n+1)\;\partial_{[j}\, p_{k]h}. \tag{2.33}$$

Putting from Eqs. (2.28) - (2.30), (2.32) and (2.33) in Eq. (5.2.15), we finally note that the projective curvature tensor vanishes identically for such choice of connection parameters. Thus, we conclude the

Theorem 2.3. A Finsler space F_n, $n > 2$, admitting a special form of the connection parameters given by Eq. (2.13), is projectively flat.

From Eq. (2.29) it is clear that the symmetry of the contracted curvature tensor H_{kh} in its indices implies and is implied by the relation

$$\partial_k p_h = \partial_h p_k, \tag{2.34}$$

which suggests a gradient form of the vector p_k:

$$p_k = \partial_k \rho, \tag{2.35}$$

for some scalar function ρ. Existence of such a function ρ ascertains the projective flatness of the space F_n, $n > 2$, having a symmetric contracted curvature tensor H_{kh}. Accordingly, from Corollary 2.1 and Theorem 2.3 there follows the

Corollary 2.2. A Finsler space F_n, $n > 2$, has connection parameters of the form Eq. (2.13) together with Eq. (2.35) iff it admits Eq. (2.10b).

§ 3. Projectively flat symmetric F_n

A symmetric Finsler space, characterized by Eqs. (7.1.1) - (7.2.3), is defined in Chapter 7. Hence, the commutation formula (4.8.3), applied to the curvature tensor H^i_{jk}, for Eqs. (4.9.15a), (7.1.1) and (7.2.1), thus yielded the identity vide Eq. (7.2.6) together with its contraction form for indices i and j gave the identity vide Eq. (7.2.7) in a symmetric F_n. If, however, the space is also projectively flat admitting Eq. (2.3), the Eq. (7.2.6) now assumes the form

$$2\,\{(n-1)\dot{x}^i H^a_{m[j|a|} - \dot{x}^a H_{am}\delta^i_{[j}\}\, G^m_{k]hl} = 0, \tag{3.1}$$ [34)]

where simplifications are carried out by means of Eqs. (4.10.15) and (7.2.7). A transvection of above equation by $\dot{x}^k$ and use of Eqs. (4.9.21a), (4.10.3), (4.10.15) and (7.2.7), further reduce it to

$$\dot{x}^a H_{am}\, G^m_{jhl} = 0, \tag{3.2}$$

for $n > 2$ and $\dot{x}^i$ being arbitrary. Next, putting from Eq. (4.9.22) in Eq. (3.2) and applying Eq. (7.2.7) again, we derive

$$(\dot{\partial}_m H) G^m_{jhl} = 0, \tag{3.3}$$

which on comparison with Eq. (7.2.7) suggests a linear relation between the curvature vectors H_m and $\dot{\partial}_m H$:

$$H_m = \chi \dot{\partial}_m H, \tag{3.4}$$

for some scalar coefficient χ. A transvection of Eq. (3.4) with $\dot{x}^m$, for Eq. (4.9.21c) and homogeneous property of *H*, determines

$$\chi = (n-1)/2. \tag{3.5}$$

Accordingly, Eq. (3.4) assumes the form

$$H_m = \{(n-1)/2\} \dot{\partial}_m H, \tag{3.6}$$

together with its directional derivatives:

$$H_{ml} = \{(n-1)/2\} \dot{\partial}_l \dot{\partial}_m H. \tag{3.7}$$

This relation establishes the symmetric property of the contracted curvature tensor H_{ml} in a projectively flat symmetric F_n, $n > 2$, implying Eqs. (2.6a) and (2.7). Consequently, Eqs. (2.2) and (2.3) reduce to Eqs. (2.8) and (2.10a) respectively. Thus, from Theorems 2.1 and 2.2 there also result the

Theorem 3.1. Asymmetric Finsler space F_n, $n > 2$, is projectively flat iff there holds Eq. (2.8).

§ 4. A projectively flat *HR-F_n*

Applying the commutation formula (4.8.3) for the vector H_k and tensor H_{kh} and noting Eqs. (8.1.1), (8.2.1) - (8.2.2) and (8.2.4a), we derive

$$\mathfrak{B}_l \dot{\partial}_m H_k = \lambda_l \dot{\partial}_m H_k + H_a G^a_{klm}, \tag{4.1a}$$

and

$$\mathfrak{B}_l \dot{\partial}_m H_{kh} = \lambda_l \dot{\partial}_m H_{kh} + H_{ah} G^a_{klm} + H_{ka} G^a_{hlm} \tag{4.1b}$$

in an $HR\text{-}F_n$, $n > 2$. Similarly, for the contracted curvature tensor H^a_{jka} we derive

$$\mathfrak{B}_l\, \dot\partial_m H^a_{jka} = \lambda_l\, \dot\partial_m H^a_{jka} - 2\, G^b_{l\,m[j} H^a_{k]ba}. \qquad (4.2)$$

Assuming $HR\text{-}F_n$, $n > 2$, projectively flat so that there holds Eq. (2.2) whose covariant derivation, for Eqs. (4.10.15), (7.2.7), (8.1.1) - (8.2.2), (4.1) and (4.2), yields

$$(n^2-1)\,\mathfrak{B}_l\, H^i_{jkh} = (n-1)\,\{\lambda_l\big(H^a_{jka}\,\delta^i_h + \dot x^i \dot\partial_h H^a_{jka}\big) - 2\dot x^i\, G^b_{l\,h[j} H^a_{k]ba}\}$$

$$-2\{\lambda_l\,(nH_{[j\,|h\,|} + H_{h\,[j} + \dot x^a \dot\partial_h H_{a[j}) + \dot x^a H_{ab}\, G^b_{l\,h[j}\}\,\delta^i_{k]}.$$

Forming its skew-symmetric part with respect to l, j, k, using Eq. (4.10.5) and symmetric properties of the tensor G^b_{hhj} in its lower indices, we derive

$$(1-n^2)\; H^m_{[jk} G^i_{l]hm} \;=\; (n-1)\lambda_{[l}\big(H^a_{jk]a}\delta^i_h + \dot x^i \dot\partial_{|\;h|}\; H^a_{jk]a}\big)$$

$$-2\lambda_{[l}\,\{nH_{j\,|h\,|} + H_{|h\,|j} + \dot x^a\; \dot\partial_{|h}\; H_{a|\,j\}}\}\,\delta^i_{k]}. \qquad (4.3)$$

Meher [89], Eq. (3.9), establishes vanishing of the contracted part of left hand terms in Eq. (4.3) with respect to indices i and h. Also, there follow

$$\dot x^i \dot\partial_i H^a_{jka} \;=\; 0,$$

and

$$\dot\partial_k H_{a\,j} - \dot\partial_j H_{a\,k} = \;\dot\partial_k\, \dot\partial_j H_a - \dot\partial_j\, \dot\partial_k H_a = 0, \qquad (4.4)$$

for the homogeneous properties of the curvature tensor and Eq. (4.9.21b) respectively. Consequently, a contraction of Eq. (4.3) with respect to indices i and h, yields

$$\lambda_{[l}\; \{n\,(n-1)\, H^a_{jk]a} - 2\,(n\, H_{jk]} \;+\; H_{k\,j]}\;)\}. \qquad (4.5)$$

As skew-symmetry implies $\lambda_{[l}\; H_{k\,j]} = -\lambda_{[l}\; H_{j\,k]}$, use of Eq. (4.10.3) reduces Eq. (4.5) to the form

$$\lambda_{[l}\; H_{jk]} = 0,$$

or

$$\lambda_l(H_{jk} - H_{kj}) + \lambda_j(H_{kl} - H_{lk}) + \lambda_k(H_{lj} - H_{jl}) = 0.$$

Its transvection with $\dot{x}^j$ and use of Eq. (4.9.21a), also yields

$$\lambda_l(\dot{x}^j H_{jk} - H_k) + \dot{x}^j \lambda_j (H_{kl} - H_{lk}) + \lambda_k (H_l - \dot{x}^j H_{jl}) = 0. \quad (4.6)$$

On the other hand, a contraction of Eq. (4.3) for i and j, and transvection by $\dot{x}^h$, for Eqs. (4.9.21a), (4.10.3), (4.10.15) and the homogeneous properties of the curvature tensor, yield

$$2\{(n^2 - n - 1) H_{[k} - \dot{x}^a H_{a[k}\} \lambda_{l]}$$

$$+ (n-1)\dot{x}^a \lambda_a (H_{lk} - H_{kl}) = 0. \quad (4.7)$$

Eliminating the term $\dot{x}^a \lambda_a (H_{kl} - H_{lk})$ from Eqs. (4.6) and (4.7), there results the identity

$$\{n H_{[k} + \dot{x}^a H_{a[k}\} \lambda_{l]} = 0,$$

for an $HR\text{-}F_n$, $n > 2$. Above identity suggests a proportionality relation between the vector fields $n H_k + \dot{x}^a H_{ak}$ and λ_k:

$$n H_k + \dot{x}^a H_{ak} = (n^2 - 1) P \lambda_k, \quad (4.8)$$

for some suitable factor P. The left hand vector being positively homogeneous of degree 1 in its directional arguments imposes the same character on P. A directional derivation of Eq. (4.8), for Eqs. (4.9.21b) and (8.2.4a), yields

$$n H_{kj} + H_{jk} + \dot{x}^a \dot{\partial}_j H_{ak} = (n^2 - 1) P_j \lambda_k, \quad (4.9)$$

where

$$P_j \equiv \dot{\partial}_j P \quad (4.10a)$$

is a vector field satisfying

$$P_j \dot{x}^j = P \quad (4.10b)$$

for the homogeneous character of P. Skew-symmetric part of Eq. (4.9) with respect to j, k, in view of Eqs. (4.10.3) and (4.4), finally determines the contracted curvature tensor H^a_{jka}:

$$H^a_{jka} = 2(n+1) P_{[j} \lambda_{k]}, \quad (4.11)$$

together with its directional derivative

$$\dot{\partial}_h H^a_{jka} = 2(n+1)\, P_{h[j}\, \lambda_{k]}, \tag{4.12}$$

where

$$P_{hj} \equiv \dot{\partial}_h P_j \quad \text{(4.13a); satisfies} \quad \dot{x}^h P_{hj} = 0, \tag{4.13b}$$

for homogeneous properties of the function P. In consequence of the relations (4.8), (4.9), (4.11) and (4.12), the tensors H^i_{jkh} and H^i_{jk} given by Eqs. (2.2) and (2.3) in a projectively flat HR- F_n, are expressible as

$$H^i_{jkh} = 2\{\delta^i_h P_{[j} + \dot{x}^i P_{h[j} + P_h \delta^i_{[j}\}\, \lambda_{k]}, \tag{4.14}$$

and

$$H^i_{jk} = 2\{\dot{x}^i P_{[j} + P\delta^i_{[j}\}\, \lambda_{k]}. \tag{4.15}$$

For Eqs. (4.9.20), (4.10b) and (4.13b), above tensors also give rise to

$$H_{kh} = n\lambda_k P_h - P_k \lambda_h - \lambda P_{kh}, \tag{4.16}$$

and

$$H_k = n P \lambda_k - \lambda P_k, \tag{4.17}$$

where

$$\lambda \equiv \dot{x}^i \lambda_i \quad \text{(4.18a); together with} \quad \dot{\partial}_k \lambda = \lambda_k \tag{4.18b}$$

for Eq. (8.2.4a). Finally, a transvection of Eq. (4.17) by $\dot{x}^k$, for Eqs. (4.9.21c), (4.10b) and (4.18), evaluates the function P:

$$P = H/\lambda, \tag{4.19}$$

and its directional derivatives

$$P_k = (\dot{\partial}_k H)/\lambda - (H/\lambda^2)\lambda_k = (1/\lambda)(\dot{\partial}_k H - P\lambda_k), \tag{4.20}$$

and

$$P_{jk} = (\dot{\partial}_j \dot{\partial}_k H)/\lambda - (2/\lambda^2)\, \lambda_{(j}\, \dot{\partial}_{k)}\, H + (2H/\lambda^3)\lambda_j\, \lambda_k. \tag{4.21}$$

Elimination of P and its directional derivatives from Eqs. (4.14) - (4.17) by means of Eqs. (4.19) - (4.21), thus, determines

$$H^i_{jkh} = (2/\lambda)\,[\{\delta^i_h + \dot{x}^i(\dot{\partial}_h - \lambda_h/\lambda)\}\, \dot{\partial}_{[j} H$$
$$+ (\dot{\partial}_h - \lambda_h/\lambda)\, H\delta^i_{[j}]\, \lambda_{k]}, \tag{4.22}$$

$$H^i_{jk} = (2/\lambda)\{\dot{x}^i\,\dot{\partial}_{[j}H + H\delta^i_{[j}]\lambda_{k]}, \qquad (4.23)$$

$$H_{kh} = \{(n+1)/\lambda\}\,\lambda_k\,(\dot{\partial}_h - \lambda_h/\lambda)\,H - \dot{\partial}_k\,\dot{\partial}_h H, \qquad (4.24)$$

and

$$H_k = \{(n+1)\,H/\lambda\}\,\lambda_k - \dot{\partial}_k H, \qquad (4.25)$$

in a projectively flat *HR*- F_n, $n > 2$.

If, however, the space under consideration were having a direction-free curvature tensor a directional derivation of Eq. (4.16), in view of Eqs. (8.2.4a), (4.13a) and (4.18b), yields

$$0 = n\,\lambda_k\,P_{jh} - P_{jk}\,\lambda_h - \lambda_j\,P_{kh} - \lambda\,\dot{\partial}_j P_{kh}.$$

On transvection with $\dot{x}^k$ the above relation, for homogeneous properties of P and its derivatives, determines

$$P_{jh} = 0, \qquad (4.26)$$

concluding invariance of the vector P_h on the directional arguments. Conversely, when Eq. (4.26) holds, the Eqs. (4.14) and (4.16) reduce to:

$$H^i_{jkh} = 2\{\delta^i_h\,P_{[j} + P_h\delta^i_{[j}\}\,\lambda_{k]}, \qquad (4.27)$$

and

$$H_{kh} = n\lambda_k\,P_h - P_k\,\lambda_h. \qquad (4.28)$$

We, therefore, conclude the

Theorem 4.1. The curvature tensor of a projectively flat *HR*-F_n, $n > 2$, would be a point function iff the vector P_j were independent of the directional arguments.

On the other hand, if the space under discussion possesses a symmetric contracted curvature tensor H_{kh} it follows from Eqs. (4.10.3) and (4.11) that the vectors P_j and λ_k satisfy

$$P_{[j}\,\lambda_{k]} = 0 \qquad (4.29) \qquad \Rightarrow \qquad P_j = Q\lambda_j, \qquad (4.30)$$

for some scalar function Q. A transvection of Eq. (4.30) by $\dot{x}^j$, for the relations (4.10b), (4.18a) and (4.19), determines

$$Q = P/\lambda = H/\lambda^2, \tag{4.31}$$

and, therefore, we have

$$P_j = (H/\lambda^2)\,\lambda_j, \tag{4.32}$$

together with

$$P_{hj} = \{(\dot{\partial}_h H)/\lambda^2 - (2H/\lambda^3)\lambda_h\}\lambda_j\,. \tag{4.33}$$

Consequently, we have

$$2\,P_{h[j}\,\lambda_{k]} = 0, \tag{4.34}$$

and the equations (4.14) - (4.16) further reduce to the forms

$$H^i_{jkh} = 2\,\delta^i_{[j}\,\lambda_{k]}\,P_h = 2(H/\lambda^2)\,\delta^i_{[j}\,\lambda_{k]}\,\lambda_h, \tag{4.35}$$

$$H^i_{jk} = 2\,\delta^i_{[j}\,\lambda_{k]}P = 2(H/\lambda)\,\delta^i_{[j}\,\lambda_{k]}, \tag{4.36}$$

$$H_{kh} = (n-1)\lambda_k\,P_h = (n-1)\,(H/\lambda^2)\,\lambda_k\,\lambda_h \tag{4.37}$$

respectively, which coincide with the corresponding ones dealt in the Section 4 of [119]. Substitution from Eq. (4.37) in Eq. (4.35) reaffirms the relation (2.8) and, therefore, Eq. (2.10b) as well. Comparison of Eqs. (4.36) and (2.10b) determines

$$\lambda_k = (\lambda/F)\,l_k\,. \tag{4.38}$$

The vector field l_k being a unit vector the magnitude of the recurrence vector λ_k is (λ/F). Thus, we have the

Theorem 4.2. If a projectively flat *HR-*F_n, $n > 2$, has a symmetric contracted curvature tensor its recurrence vector is given by Eq. (4.38).

Also, in the light of Theorem 3.1. of [135], where an *HR-*F_n, $n > 2$, admitting an infinitesimal transformation generated by a concircular vector field necessarily has a symmetric contracted curvature tensor. Further, it is seen (cf. Corollary 3.1 of [135]) that such a space is projectively flat.

§ 5. Some more properties of the functions P and λ_j in a projecttively flat *HR-*F_n

Eliminating $P\lambda_k$ from Eqs. (4.17) and (4.20), we obtain

$$\lambda\, P_k = (n\, \dot{\partial}_k H - H_k) / (n+1). \qquad (5.1)$$

In view of commutation rule in Eq. (4.8.3), applied for the scalar curvature H, and Eq. (8.2.4a), a directional derivation of Eq. (8.2.2c) establishes the recurrent property of the derivative $\dot{\partial}_k H$, while Eq. (8.2.2b) states the same character of the curvature vector H_k of $HR\text{-}F_n$. This implies the recurrent character of the vector in Eq. (5.1):

$$\mathfrak{B}_j (\lambda P_k) = \lambda_j (\lambda P_k) \quad (5.2); \text{ or } \quad \mathfrak{B}_j P_k = \nu_j P_k, \qquad (5.3)$$

where

$$\nu_j \equiv \lambda_j - (\mathfrak{B}_j \lambda) / \lambda \qquad (5.4)$$

is a covariant vector field defined in terms of the vector λ_j. A transvection of Eq. (5.3) by $\dot{x}^k$, for Eq. (4.10b) and invariant nature of $\dot{x}^k$ under the covariant differentiation, also yield

$$\mathfrak{B}_j P = \nu_j P. \qquad (5.5)$$

Equations (5.3) and (5.5) express the recurrent character of the functions P_k and P respectively for the recurrence vector ν_j. Also, a covariant differentiation of Eq. (4.17), for Eqs. (8.2.2b), (5.2) and (4.17) itself, establishes the recurrent property of the vector $P\lambda_k$:

$$\mathfrak{B}_j (P\, \lambda_k) = \lambda_j (P \lambda_k), \qquad (5.6)$$

or, by Eq. (5.5)

$$\mathfrak{B}_j \lambda_k = \{\lambda_j - (\mathfrak{B}_j P) / P\} \lambda_k = (\lambda_j - \nu_j) \lambda_k = \mu_j\, \lambda_k, \qquad (5.7)$$

where

$$\mu_j \equiv \lambda_j - \nu_j = (\mathfrak{B}_j \lambda) / \lambda, \qquad (5.8)$$

by Eq. (5.4). Forming its directional derivative, applying the commutation formula vide Eq. (4.8.3) for the scalar function λ, and using Eq. (4.18b), we obtain

$$\dot{\partial}_k \mu_j = (\mathfrak{B}_j \lambda_k) / \lambda - (\mathfrak{B}_j \lambda) \lambda_k / \lambda^2 = \mu_j \lambda_k / \lambda - \mu_j \lambda_k / \lambda = 0, \qquad (5.9)$$

by Eqs. (5.7) and (5.8). For Eqs. (8.2.4a) and (5.9), there also follows

$$\dot{\partial}_k \nu_j = 0, \tag{5.10}$$

from Eq. (5.8). Thus, like the recurrence vector λ_j the vectors μ_j and ν_j are also independent of the directional arguments.

Now we are in position to look for the recurrent property of the second order directional derivatives of the function P. Differentiating Eq. (5.3) partially with respect to $\dot{x}^h$, applying the commutation rule vide Eq. (4.8.3) for the vector field P_k, and using Eqs. (4.13a) and (5.10), we derive

$$\mathfrak{B}_j P_{hk} = P_a G^a_{jkh} + \nu_j P_{hk}. \tag{5.11}$$

To evaluate $P_a G^a_{jkh}$ we, first, apply the commutation formula (4.8.3) to the recurrence vector λ_h, and use Eqs. (8.2.4a), (5.7) and (5.9) to get

$$\lambda_a G^a_{jkh} = 0. \tag{5.12}$$

Next, transvecting Eq. (4.17) by G^k_{jhl} and recalling Eq. (7.2.7) [35] and (5.12) we immediately deduce

$$P_k G^k_{jhl} = 0. \tag{5.13}$$

Hence, Eq. (5.11), finally, reduces to

$$\mathfrak{B}_j P_{hk} = \nu_j P_{hk}, \tag{5.14}$$

establishing the recurrent property of the tensor P_{hk}. Further directional derivatives of Eq. (5.14) would similarly yield

$$(\mathfrak{B}_j - \nu_j) P_{lhk} = 2 P_{a(k} G^a_{h)lj}, \tag{5.15}$$

where

$$P_{lhk} = \dot{\partial}_l P_{hk}. \tag{5.16}$$

To evaluate the terms on the right side of Eq. (5.15), we differentiate Eq. (5.13) with respect to $\dot{x}^a$ and use (4.13a):

$$P_{ak} G^k_{jhl} + P_k G^k_{jhla} = 0, \qquad G^k_{jhla} \equiv \dot{\partial}_a G^k_{jhl} \quad .$$

Interchanging the indices a and h, and using the symmetry of the tensors P_{ak}, G^k_{jhl} and G^k_{jhla} in their (lower) indices, we derive

$$2 P_{k(a} G^k_{h)lj} = -2 P_k G^k_{ahlj}, \tag{5.17}$$

establishing non-vanishing character of the right side of Eq. (5.15). This prevents the tensor P_{lhk} to be recurrent in general. We, therefore, conclude the

Theorem 5.1. The recurrent vector field λ_k as well as the factor of proportionality P (together with its directional derivatives at least up to second order) are recurrent in a projectively flat HR-F_n, $n > 2$, and the corresponding recurrence vectors μ_j and ν_j (like the vector λ_j) are at most point functions in the space.

In the following, we derive an explicit relation between the vector fields λ_k and P_k. Transvecting Eq. (4.15) by l_i, and putting from Eqs. (4.9.23) and (2.11), we get

$$(F P_{[j} + P l_{[j}) \lambda_{k]} = 0. \tag{5.18}$$

Its further transvection by $\dot{x}^k$, for Eqs. (4.11.6), (4.10b) and (4.18a), yields the desired result:

$$P_j / P + l_j / F = 2\lambda_j / \lambda, \tag{5.19a}$$

or,

$$(\dot{\partial}_j P)/P + (\dot{\partial}_j F)/F = 2(\dot{\partial}_j \lambda)/\lambda, \tag{5.19b}$$

for Eqs. (4.2.4), (4.10a) and (4.18b). Above differential equation is term wise integrable with respect to $\dot{x}^j$ having the solution:

$$\ln P + \ln F = 2 \ln \lambda + \ln \varphi \;\Rightarrow\; PF/\lambda^2 = \varphi, \tag{5.20}$$

where φ (x^k), being the parameter of integration with respect to the directional arguments, may be an arbitrary scalar point function. Putting from Eq. (4.19) in Eq. (5.20), we note that the quotient function

$$\varphi \equiv FH/\lambda^3 \tag{5.21}$$

is, therefore, independent of the directional arguments in a projectively flat $HR\text{-}F_n$. Rewriting Eq. (5.19b) as

$$P_j = P\dot{\partial}_j(\ln \lambda^2 - \ln F) = P\dot{\partial}_j f, \tag{5.22}$$

where

$$f \equiv \ln(\lambda^2/f). \tag{5.23}$$

Accordingly, the directional derivative of Eq. (5.22), for Eqs. (4.10a) and (5.22) itself, assumes the form

$$P_{jk} = P\{(\dot{\partial}_j f)(\dot{\partial}_k f) + \dot{\partial}_j \dot{\partial}_k f\}, \tag{5.24}$$

Now raising a conjecture if the space has a direction-free curvature tensor there results the differential equation

$$\dot{\partial}_j \dot{\partial}_k f + (\dot{\partial}_j f)(\dot{\partial}_k f) = 0, \tag{5.25}$$

from Eq. (5.24) and the Theorem 4.1. It is evident from Eq. (5.22) that the derivative $\dot{\partial}_j f$ is a positively homogeneous function of degree – 1 in $\dot{x}^i$'s, and so is the function f of degree zero. Accordingly, a transvection of Eq. (5.25) by $\dot{x}^j$ deduces

$$\dot{\partial}_k f = 0, \tag{5.26}$$

that causes vanishing of P_j as per Eq. (5.22). Therefore, it follows from Eq. (4.27) that the curvature tensor H^i_{jkh} vanishes identically leading to a contradiction for an $HR\text{-}F_n$. Thus, above conjecture is impossible and we conclude the

Theorem 5.2. A projectively flat $HR\text{-}F_n$, $n > 2$, cannot admit a direction-free curvature tensor.

In the rest of this section we study the properties of vector fields λ_k and P_k in a projectively flat $HR\text{-}F_n$, $n > 2$, admitting an affine motion. As $\dot{x}^i$ remains invariant under the Lie derivation the integrability conditions in Eq. (6.8.5) for an affine motion also imply

$$\pounds H = 0, \tag{5.27}$$

for Eqs. (6.8.10). Taking Lie derivative of Eq. (8.2.2c), applying the commutation rule vide Eq. (6.6.7) for the scalar curvature *H*, and noting Eqs. (6.8.2) and (5.27), we also derive

$$£\,\lambda_l = 0 \qquad (5.28); \quad \text{together with } £\,\lambda = 0, \qquad (5.29)$$

as in Eq. (7.7.1). Consequently, there follows Lie invariance of the function *P* as well:

$$£\,P = 0, \qquad (5.30)$$

from Eqs. (4.19) and (5.27). Successive derivations of (5.30) with respect to directional coordinates, in view of Eqs. (6.6.1) and (4.10a), also yield

$$£\,P_j = 0, \quad £\,P_{jk} = 0, \text{ etc.} \qquad (5.31)$$

We, therefore, have the

Theorem 5.3. The recurrence vector field λ_k and the factor of proportionality *P* (together with its directional derivatives of any order) are Lie-invariants in a projectively flat *HR-F*$_n$, $n > 2$, admitting an affine motion.

§ 6. Normal projective curvature tensor in a projectively flat *HR-F*$_n$

The normal projective curvature tensor N^i_{jkh} introduced by Yano [234], p.196, is related to Berwald's curvature tensor H^i_{jkh} by [141], Eq. (3.2):

$$N^i_{jkh} = H^i_{jkh} - (\dot{x}^i\, \dot{\partial}_h\, H^a_{jka}\,)/(n+1). \qquad (6.1)$$

Putting from Eqs. (4.12) and (4.14) in above equation an expression for N^i_{jkh} is derived in terms of the vectors λ_k and P_k:

$$N^i_{jkh} = 2\{\delta^i_h\, P_{[j} + P_h\, \delta^i_{[j}\}\,\lambda_{k]}, \qquad (6.2)$$

which (unlike H^i_{jkh}) does not involve the second order directional derivatives of the function *P*. Furthermore, the tensors

$$N_{kh} \equiv N^i_{ikh} \qquad \text{and} \quad V_{kh} \equiv (n\,N_{kh} + N_{hk})/(1-n^2)$$

are then expressible as

$$N_{kh} = n\lambda_k P_h - P_k \lambda_h \quad (6.3); \text{ and } \quad V_{kh} = -\lambda_k P_h. \qquad (6.4)$$

Taking covariant derivatives of Eqs. (6.2) - (6.4), and using Eqs. (5.3), (5.7) and (5.8) it can be easily verified that the normal projective curvature tensor N^i_{jkh} along with its contracted forms also possess recurrent properties with respect to the recurrence vector λ_l. Thus, we have the

Theorem 6.1. The projective flatness of an HR-F_n, $n > 2$, induces the recurrent character on its normal projective curvature tensors:

$$(\mathfrak{B}_l - \lambda_l)\, N^i_{jkh} = (\mathfrak{B}_l - \lambda_l)\, N_{kh} = (\mathfrak{B}_l - \lambda_l)\, V_{kh} = 0. \quad (6.5)$$

Foot-notes:

34) The index a within two solidi contributes nothing to the skew-symmetric part of the object.

35) Eqs. (7.2.5) and (7.2.7) holding in a symmetric F_n also hold in an HR-F_n (cf. Eq. (8.2.16)).

CHAPTER 12

ON FINSLER SPACES WITH CONCIRCULAR TRANSFORMATIONS – I

§ 1. Introduction

Concircular transformations were introduced by Yano [233] in Riemannian geometry. Later, these were extended by Takano [220] - [222] to affine geometry with recurrent curvature and were further studied by Okumura [161] in different types of Riemannian manifolds. Takano [220-V] also discussed various alternative forms for the covariant derivative of the generator of an infinitesimal transformation; which were, later, extended by Misra et al. [135] to Finslerian manifolds of recurrent curvature confining to the discussion of concircular form of the generator only. Later, it was hastily deduced by Pandey [175] that a recurrent Finsler manifold cannot admit a concircular transformation (also, cf. [121], Theorem 3.2); whereas a symmetric Finsler space with such transformation reduces to a symmetric Riemannian space [179]. Their particular cases: contra, concurrent and special concircular transformations defining projective motions were also studied by the author [113] in a recurrent Finsler manifold. The discussion of concircular infinitesimal transformation (here onwards abbreviated as c.i.t.) in different types of Finslerian spaces: flat, isotropic, with constant sectional curvature and the symmetric ones is continued in the present chapter.

§ 2. F_n with a concircular vector field

The discussion of Sub-section 9.2 of Chapter 5 is continued here to derive some more results when a Finsler space F_n is transformed under the infinitesimal change vide Eq. (5.4.1) generated by a concircular vector field. The covariant derivation of Eq. (5.9.18) with respect to x^j, for Eqs. (5.9.1e), (5.9.11a) and (5.9.18) itself, gives

$$(\mathfrak{B}_j H_{kh})\, v^h + \rho\, H_{kj}$$

$$= (n-1)\{2\rho_{(j}\, \mu_{k)} + \rho\,(\mathfrak{B}_j \mu_k - \mu_j\, \mu_k) - \mathfrak{B}_j \rho_k\}. \quad (2.1)$$

In view of Eqs. (5.9.2) and (5.9.11b), RHS of Eq. (2.1) is symmetric in j, k. Therefore, its skew-symmetric part in j, k, for Eq. (4.10.3), yields

$$2\,(\mathfrak{B}_{[j}\,H_{k]h})\,v^h + \rho\, H^a_{jka} = 0. \tag{2.2}$$

Now we define a tensor

$$M^i_{jkh} \equiv H^i_{jkh} - 2\delta^i_{[j} H_{k]h}/(n-1), \tag{2.3}$$

and consider its contracted forms and other associated tensors in analogy with those defined by Eqs. (4.9.15), (4.9.20), (4.9.21), and (4.10.3):

$$M^i_{jk} \equiv M^i_{jkh}\dot{x}^h = H^i_{jk} - 2\delta^i_{[j} H_{k]}/(n-1), \tag{2.4}$$

$$\dot{\partial}_h M^i_{jk} = M^i_{jkh}, \tag{2.5}$$

$$M^i_j \equiv M^i_{jk}\dot{x}^k = H^i_j - H\,\delta^i_j + (\dot{x}^i H_j)/(n-1), \tag{2.6}$$

$$\left.\begin{array}{l} \text{a) } M_{kh} \equiv M^i_{ikh} = 0, \quad \text{b) } M_k \equiv M^i_{ik} = 0, \\ \text{c) } M^i_i = M_k \dot{x}^k = 0, \text{ d) } M^i_{jki} = \{(n-2)/(n-1)\}H^i_{jki}. \end{array}\right\} \tag{2.7}$$

Also, from Eqs. (4.10.3), (4.10.1a), (4.10.5) and (2.3), there result the generalized Bianchi identities satisfied by M^i_{jkh} :

$$M^i_{[jkh]} = \delta^i_{[j} H^a_{kh]a}/(n-1), \tag{2.8}$$

$$\mathfrak{B}_{[l}\, M^i_{jk]h} = G^i_{hm[l}\, H^m_{kj]} + 2\delta^i_{[l}\mathfrak{B}_j H_{k]h}/(n-1). \tag{2.9}$$

Transvecting Eq. (2.9) by v^h, and using Eq. (2.2) we also have

$$v^h\,\mathfrak{B}_{[l}\, M^i_{jk]h} = v^h G^i_{hm[l} H^m_{kj]} - \rho\delta^i_{[l} H^a_{jk]a}/(n-1). \tag{2.10}$$

Rewriting Eq. (5.9.15a) in view of Eq. (2.3):

$$M^i_{jkh} v^h = 0, \tag{2.11}$$

forming its covariant derivative, and using Eqs. (5.9.1e) and (2.11) itself, there follows

$$v^h\,\mathfrak{B}_l\, M^i_{jkh} + \rho M^i_{jkl} = 0. \tag{2.12}$$

Its skew-symmetric part with respect to the indices l, j, k, for Eqs. (2.8) and (2.10), finally yields

$$v^h G^i_{hm[l} H^m_{kj]} = 0. \tag{2.13}$$

Using Eq. (5.9.1e) and various properties of the functions ρ and μ_j mentioned above, now we are in position to express Lie derivatives of the metric tensor and the connection parameters in an F_n admitting a c.i.t. Writing the formula (6.2.16b) for the metric tensor, putting from Eq. (5.9.1e) and simplifying by means of Eqs. (3.9.1), (4.2.9), (5.10.4) and $v_j \equiv g_{ij} v^i$, we thus, obtain

$$\pounds\, g_{ij} = (\mathfrak{B}_k\, g_{ij} + 2\,\mu\, C_{ijk})\, v^k + 2\{\rho\, g_{ij} + \mu_{(i}\, v_{j)}\}. \qquad (2.14)$$

In particular, for a Berwald space admitting a c.i.t. with an additional condition $\mu_i = 0$, the Eq. (2.14) reduces to the form of Eq. (6.11.1). Thus, we conclude the

Theorem 2.1. Every special c.i.t. defines a conformal motion in a Berwald space.

As a particular case of this theorem, we also have the

Corollary 2.1. Every concurrent (respectively contra) infinitesimal transformation defines a homothetic motion (resp. motion) in a Berwald space.

Similarly, substitutions from Eqs. (5.9.1e) and (5.9.7) in Eq. (6.3.7), for Eqs. (4.10.15) and (5.10.4), determine

$$\pounds\, G^i_{jk} = \rho\, \delta^i_j\, \mu_k + \rho_j\, \delta^i_k + (\mathfrak{B}_j \mu_k + \mu_j \mu_k)\, v^i$$
$$+ (H^i_{hjk} + \mu G^i_{hjk})\, v^h. \qquad (2.15)$$

This result will be more useful in the study of flat manifolds admitting a c.i.t. dealt in the next Section.

§ 3. Flat manifolds admitting a c.i.t.

In a flat manifold the curvature tensor and, therefore, its associated tensors and contracted forms vanish identically. Thus, from Eq. (5.9.18), there follows

$$\rho\, \mu_k = \rho_k \quad \Rightarrow \quad \mu_k = \rho_k / \rho = \mathfrak{B}_k \ln \rho, \qquad (3.1)$$

in a flat Finsler manifold admitting a c.i.t. For ρ and ρ_k being point functions, Eq. (3.1) establishes independence of μ_k over the directional arguments. Hence, the commutation formula (4.8.3), when applied for the vector v^i (satisfying Eq. (5.9.1e)), yields Eq. (5.9.12) and therefore

$$\pounds\,(G^i_{jkh}v^h) = 0 \qquad (3.2)$$

also. Differentiating Eq. (3.1) covariantly with respect to x^j, and putting from Eqs. (5.9.11a) and (3.1) itself, we obtain

$$\mathfrak{B}_j\,\mu_k + \mu_j\mu_k = (\mathfrak{B}_j\,\rho_k)/\rho. \qquad (3.3)$$

In view of Eqs. (5.9.12), (3.1) and (3.3), the formula (2.15) assumes the form

$$\pounds\,G^i_{jk} = 2\rho_{(j}\delta^i_{k)} + v^i\,(\mathfrak{B}_j\,\rho_k)/\rho \qquad (3.4)$$

in a flat Finsler manifold admitting a c.i.t. Transvecting it by $\dot{x}^j$, and using Eqs. (4.6.2a) and (6.2.9b), we have

$$\pounds\,G^i_k = \dot{x}^i\,\rho_k + \dot{x}^j\rho_j\delta^i_k + v^i\,\dot{x}^j(\mathfrak{B}_j\,\rho_k)/\rho. \qquad (3.5)$$

Its further transvection by the tensor G^a_{ihj}, for Eq. (4.10.15) and (5.9.12), also yields

$$G^a_{ihj}\,\pounds\,G^i_k = \dot{x}^l\,\rho_l G^a_{khj}. \qquad (3.6)$$

The right hand tensor being symmetric in *j*, *k* the skew-symmetric part of Eq. (3.6) with respect to these indices vanishes:

$$2\,G^a_{ih[j}\,\pounds\,G^i_{k]} = 0. \qquad (3.7)$$

Accordingly, the formula (6.8.9) reduces to

$$2\mathfrak{B}_{[j}\,\pounds\,G^i_{k]h} = 0 \qquad (3.8)$$

in such a manifold.

Differentiating Eq. (3.4) partially w.r.t. $\dot{x}^h$, applying the commutation formulae vide Eqs. (4.8.3) and (6.6.12), and noting the independence of ρ, ρ_j and v^i on $\dot{x}^h$'s, we obtain

$$\pounds\,G^i_{hjk} = -(1/\rho)\,v^i\,\rho_a\,G^a_{jkh} \qquad \Rightarrow \qquad \pounds\,G^i_{hji} = 0, \qquad (3.9)$$

by Eq. (5.9.12). This implies the equivalence of Lie derivatives of the normal projective connection parameters given by Eq. (10.4.1) (cf. [234], p. 196) and Berwald's connection parameters G^i_{jk} :

$$\pounds \overset{(n)}{\Pi}{}^i_{jk} = \pounds\, G^i_{jk} \text{ } ^{36)} \tag{3.10}$$

w.r.t. a c.i.t. in a flat Finsler F_n . A projective motion defined by η is characterized by ([234], p. 200):

$$\pounds \overset{(n)}{\Pi}{}^i_{jk} = 2\delta^i_{(j} p_{k)}, \tag{3.11}$$

where the functions p_k are given by Eq. (5.1.9a). Comparing Eqs. (3.10) and (3.11) we, therefore, conclude the

Theorem 3.1. A projective motion defined by a c.i.t. in a flat F_n is necessarily of a restricted form:

$$\pounds\, G^i_{jk} = 2\delta^i_{(j} p_{k)}. \tag{3.12}$$

For compatibility of Eqs. (3.4) with (3.12) we must, therefore, have

$$\text{a)} \quad \rho_j = p_j, \qquad \text{b)} \quad \mathfrak{B}_j\, \rho_k = 0 \tag{3.13}$$

for a projective motion defined by a c.i.t. in a flat F_n . Conversely, when Eqs. (3.13) hold the Eq. (3.4) reduces to the form of Eq. (3.12) implying a c.i.t. to define a projective motion. We, therefore, have the

Theorem 3.2. A c.i.t. defines a projective motion in a flat Finsler space iff there hold Eqs. (3.13).

§ 4. Isotropic Finsler space with a concircular vector field

Defining a vector filed

$$h^i_j \equiv \delta^i_j - l^i\, l_j, \tag{4.1}$$

we derive

$$\text{a)}\ h^i_i \equiv n-1, \quad \text{b)}\ h^i_j\, y_i \equiv 0, \quad \text{c)}\ 2\{ h^i_{[j} - \delta^i_{[j} \} y_{k]} = 0, \tag{4.2}$$

and

$$\dot{\partial}_j h^i_k = -\left(h^i_j l_k + l^i h^a_j g_{ka}\right)/ F = -\left(\delta^i_j l_k + l^i g_{jk} - 2 l^i l_j l_k\right)/ F, \tag{4.3}$$

from Eqs. (4.2.3), (4.2.4), (4.2.5), (4.2.7), (4.2.8) and (4.11.8). Hence, the condition in Eq. (4.11.4) for an isotopic F_n, $n > 2$, is rewritten as

$$H^{i}_{jk} = 2h^{i}_{[j}\left(F^{2}\dot{\partial}_{k]}R/3 + Ry_{k]}\right). \tag{4.4}$$

When contracted with respect to i and j, for Eqs. (4.9.20b), (4.1), (4.2) and homogeneous properties of R, it yields

$$H_{k} = \{(n-2)/3\}\, F^{2}\dot{\partial}_{k}R + (n-1)\, R\, y_{k}. \tag{4.5}$$

Its directional derivation, for Eqs. (4.2.4), (4.2.7a), (4.9.21b) and (4.11.8) also determines

$$H_{kh} = \{(n-2)/3\}(2\, y_{h}\, \dot{\partial}_{k}R + F^{2}\dot{\partial}_{k}\dot{\partial}_{h}R)$$

$$+ (n-1)(\, y_{k}\dot{\partial}_{h}R + Rg_{kh}). \tag{4.6}$$

Hence, Eq. (4.10.3) yields

$$H^{a}_{jka} = -\ \{2(n+1)/3\}\ y_{[j}\ \dot{\partial}_{k]}R). \tag{4.7}$$

Putting from Eqs. (4.4) and (4.5) in Eq. (2.4), and using Eq. (4.2c), we derive

$$M^{i}_{jk} = (2/3)F^{2}\{\, h^{i}_{[j} - (n-2)(n-1)^{-1}\ \delta^{i}_{[j}\}\, \dot{\partial}_{k]}R$$

$$= (2/3)F^{2}\{(n-1)^{-1}\ \delta^{i}_{[j} - l^{i}l_{[j}\}\, \dot{\partial}_{k]}R, \tag{4.8}$$

by Eq. (4.1). On differentiation with respect to $\dot{x}^{h}$, for Eqs. (4.2.4), (4.2.7a), (4.11.8) and (2.5), yields

$$M^{i}_{jkh} = (2/3)[\{\, 2(n-1)^{-1}\, y_{h}\ \delta^{i}_{[j} - \delta^{i}_{h}\ y_{[j} - \dot{x}^{i}\ g_{h[j}\}\, \dot{\partial}_{k]}R$$

$$+ \{F^{2}(n-1)^{-1}\ \delta^{i}_{[j} - \dot{x}^{i}\, y_{[j}\}\, \dot{\partial}_{k]}\dot{\partial}_{h}R]. \tag{4.9}$$

On contraction with respect to the indices i and h, above equation also yields

$$M^{a}_{jka} = -\,(2/3)\{\,(n+1)(n-2)/(n-1)\}\, y_{[j}\ \dot{\partial}_{k]}R, \tag{4.10}$$

for R being positively homogeneous function of degree zero in $\dot{x}^{i}$'s. This equation could be directly derived from Eqs. (2.7d) and (4.7) as well. In a space admitting a c.i.t. the Eq. (2.11), for Eq. (4.9), reduces to

$$\{2\alpha(n-1)^{-1}\delta^i_{[j} - v^i y_{[j} - \dot{x}^i v_{[j}\}\dot{\partial}_{k]}R$$

$$+ \{F^2(n-1)^{-1}\delta^i_{[j} - \dot{x}^i y_{[j}\}\dot{\partial}_{k]}\dot{\partial}_h R v^h = 0, \qquad (4.11)$$

where $\alpha \equiv y_h v^h$ is a scalar function.

§ 5. F_n with a constant sectional curvature

We consider a Finsler space F_n, $n > 2$, of constant sectional curvature satisfying Eq. (4.11.7). Its directional derivative with respect to $\dot{x}^h$ yields back the curvature tensor:

$$H^i_{jkh} = 2R\,\delta^i_{[j}\,g_{k]h}, \qquad (5.1)$$

together with its contracted form

$$H_{kh} = (n-1)\,R\,g_{kh}, \qquad (5.2)$$

where Eqs. (4.2.7a), (4.9.15a) and (4.9.20a) are used. It is easy to verify from Eqs. (2.3), (5.1) and (5.2) that the tensor M^i_{jkh} vanishes identically giving

$$H^i_{jkh} = \{2/(n-1)\}\,\delta^i_{[j}\,H_{k]h}, \qquad (5.3)$$

and, therefore, there also holds Eq. (5.9.15a) in such a space. Thus, we conclude the

Theorem 5.1. The condition in Eq. (5.9.15a) is a necessary consequence in either cases: F_n, $n > 2$, is of constant sectional curvature or it admits a c.i.t.

Let σ and f be two arbitrary scalar point functions determining the associate form of the generator v^i of infinitesimal transformation η:

$$v_h = \sigma f_h \equiv \sigma \mathfrak{B}_h f, \qquad (5.4)$$

so that

$$\mathfrak{B}_k v_h = (\mathfrak{B}_k \sigma) f_h + \sigma \mathfrak{B}_k f_h = \{(\mathfrak{B}_k \sigma)/\sigma\} v_h + \sigma \mathfrak{B}_k f_h. \qquad (5.5)$$

Since

$$\dot{\partial}_a f_h = \dot{\partial}_a \mathfrak{B}_h f = \mathfrak{B}_h \dot{\partial}_a f = 0,$$

for f being a point function, and the commutation formula (4.7.14), when applied to the vector f_h, yields

$$2\,\mathfrak{B}_{[j}\,\mathfrak{B}_{k]}\,f_h = -\,H^{a}_{jkh}\,f_a = -2R\,f_{[j}g_{k]h} \tag{5.6}$$

in a space F_n, $n > 2$, with constant sectional curvature satisfying Eq. (5.1). The differential equation (5.6) has a solution

$$\mathfrak{B}_k\,f_h = -R\,f\,g_{kh}, \tag{5.7}$$

as the skew-symmetric part of the covariant derivative of above equation:

$$\mathfrak{B}_j\,\mathfrak{B}_k\,f_h = -R\,\{f_j\,g_{kh} - 2f\,\dot{x}^a\,\nabla_a C_{jkh}\},$$

w.r.t.the indices j, k assumes the form Eq. (5.6), where Eq. (4.6.8) is also used. Putting from Eq. (5.7) in Eq. (5.5) we, thus, obtain

$$\mathfrak{B}_k\,v_h = -(R\,\sigma f)\,g_{kh} + (\mathfrak{B}_k \ln \sigma)\,v_h. \tag{5.8}$$

ln σ being a point function the coefficient of v_h in above equation satisfies Eq. (5.9.2). Comparing Eqs. (5.8) with (5.9.1e) we, thus, have

Theorem 5.2. The associate vector of the generator of η is always in concircular form in a space F_n, $n > 2$, of constant sectional curvature.

Transvecting Eq. (5.8) by g^{ih} and using $v^i \equiv g^{ih}v_h$, we derive

$$\mathfrak{B}_k\,v^i = (\mathfrak{B}_k\,g^{ih})\,v_h - (R\,\sigma f)\,\delta^i_k + (\mathfrak{B}_k \ln \sigma)\,v^i. \tag{5.9}$$

As $\mathfrak{B}_k g^{ih} \neq 0$, in general, Eq. (5.9) cannot be reduced to Eq. (5.9.1e). However, in a Berwald space having vanishing covariant derivative of the metric tensor (cf. Eq. (4.6.5), the equation (5.9) does take the form of Eq. (5.9.1e). Thus, we have the

Theorem 5.3. A Finsler space of constant sectional curvature does not admit, in general, a c.i.t. but it does so if it is also a Berwald space.

In the rest of this section we consider a Berwald space F_n, $n > 2$, of constant sectional curvature. Following [189], Eqs. (4.6.9a) and

(4.6.9c), the Berwald's curvature tensor then coincides with that of Cartan.

$$\text{a)}\ H_{jkhl} \equiv H^{i}_{jkh} g_{il} = K_{jkhl}, \quad \text{b)}\ H_{jk[hl]} = R_{jkhl}. \tag{5.10}$$

Transvecting Eq. (5.1) by g_{il} we have

$$H_{jkhl} = K_{jkhl} = 2R g_{l[j} g_{k]h}, \tag{5.11a}$$

so that Eq. (5.10b) reduces to

$$R_{jkhl} = (R/2)\,(g_{lj} g_{kh} - g_{lk} g_{jh} - g_{hj} g_{kl} + g_{hk} g_{jl})$$

$$= 2R g_{l[j} g_{k]h} = H_{jkhl}, \tag{5.11b}$$

by Eq. (5.11a), in the space under consideration. Consequently, Eq. (4.2.14) of [189] yields

$$\text{a)}\ C_{ijm} H^{m}_{kh} = 0, \quad \text{or} \quad \text{b)}\ C_{ij[k} y_{h]} = 0, \tag{5.12}$$

for Eq. (4.11.7). A transvection of Eq. (5.12b) by $\dot{x}^k$, for Eqs. (4.2.3) and (4.2.9), determines

$$C_{ijh} \equiv 0 \tag{5.13}$$

reducing F_n into a Riemannian space. Thus, we have the

Theorem 5.4. A Berwald space of constant sectional curvature is Riemannian.

§ 6. Symmetric F_n with a concircular vector field

A symmetric F_n is defined by Eq. (7.1.1) whose successive transvections, in view of Eqs. (4.9.15), (4.9.20) and (4.9.21), also cause vanishing of covariant derivatives of other associated and contracted forms of the Berwald's curvature tensor as given by Eqs. (7.2.1) - (7.2.3). If the space admits a c.i.t. there hold Eqs. (5.9.15a), (5.9.17) and (5.9.18). A covariant derivation of Eq. (5.9.15a), for Eqs. (5.9.1e), (5.9.15a), (7.1.1) and (7.2.1), yields Eq. (5.3) from which there also results

$$H^{i}_{jki} = 0, \qquad \text{if } n > 2, \tag{6.1}$$

establishing symmetry of the contracted curvature tensor H_{kh}. It causes vanishing of the projective curvature tensor given by Eq. (5.2.15) for Eq. (5.3). Thus, analogous to the Corollary 3.1 of [135], we also have the following result.

Theorem 6.1. A symmetric Finsler space F_n, $n > 2$, admitting a c.i.t. is projectively flat.

Berwald's associate curvature tensor, defined by Eq. (5.10a), is evaluated from Eq. (5.3):

$$H_{jkhl} = \{2/(n-1)\}\, g_{l[j}\, H_{k]h} \tag{6.2}$$

in a symmetric F_n, $n > 2$, admitting a c.i.t. Forming symmetric part of Eq. (6.2) in h, l and transvecting the resulting equation by $\dot{x}^l$, we obtain

$$H_{jk(hl)}\, \dot{x}^l = \{ g_{h[j}\, H_{k]} + y_{[j}\, H_{k]h} \}/(n-1), \tag{6.3}$$

for Eqs. (4.2.2) and (4.9.21a). The left member of this identity vanishes in view of Eq. (4.2.11) of [189]. A further transvection of Eq. (6.3) by $\dot{x}^k$, for Eqs. (4.2.2), (4.2.3), (4.9.21a) and symmetry of H_{kh}, determines

$$H_{jh} = \{(n-1)\, H g_{jh} + 2 y_{[j}\, H_{h]}\}/F^2. \tag{6.4}$$

The last term on the right side of this relation being not symmetric in j, h must, therefore, vanish identically:

$$y_{[j}\, H_{h]} = 0,\text{[37)]} \tag{6.5}$$

reducing Eq. (6.4) as

$$H_{jh} = (n-1)\,(H/F^2)\, g_{jh}. \tag{6.7}$$

Accordingly, Eq. (5.3) reduces to

$$H^i_{jkh} = (2H/F^2)\, \delta^i_{[j}\, g_{k]h}, \tag{6.8}$$

which does not coincide with Eq. (5.1) unless there holds Eq. (4.11.5). Assuming this condition, that holds only in a space F_n, $n > 2$, of constant sectional curvature R, Pandey [177] hastily concludes Eq. (5.1) from Eq. (6.8) and claims the reduction of a symmetric F_n to a space of constant sectional curvature. Indeed, we have proved the

Theorem 6.2. A symmetric F_n, $n > 2$, with a c.i.t necessarily admits Eqs. (5.3) and (6.8).

For Eqs. (4.2.2) and (4.9.21a), the equation (6.7) also implies

$$H_{j} = (n-1)(H / F^2)\, y_{j}. \tag{6.9}$$

A covariant derivation of Eq. (6.7), for Eqs. (4.6.7) and (7.2.3), establishes Eq. (4.6.8) reducing the symmetric F_n to a Berwald space. Thus,

Theorem 6.3. A symmetric F_n, $n > 2$, with c.i.t. is a Berwald space.

We know that a Berwald space F_n , $n > 2$, of constant sectional curvature is symmetric ([108], Theorem 2.3) and is Riemannian in view of Theorem 5.4. Hence, Theorem 2.3 of [108] is relevant in Riemannian case only. On the other hand, a covariant derivation of Eq. (5.1), for Eq. (7.1.1) and constant R, implies

$$\delta^i_j\, \mathfrak{B}_l\, g_{kh} - \delta^i_k\, \mathfrak{B}_l\, g_{jh} = 0,$$

or, on contracton w.r.t. indices i and j,

$$\mathfrak{B}_l\, g_{kh} = 0. \tag{6.10}$$

This, for Eqs. (4.6.5) and (4.6.8) establishes that a symmetric F_n, $n > 2$, of constant sectional curvature is a Berwald space. Thus, we have the

Theorem 6.4. An F_n , $n > 2$, of constant sectional curvature is a symmetric Riemannian space iff it is a Berwald space.

Consequently, Theorem 2.2 of [108] is also relevant for a Riemannian case only where the condition vide Eq. (7.2.6) is obviously satisfied for G^i_{jkh} being zero. Furthermore, it follows from Eqs. (5.1) and (6.10) that a Berwald space of constant sectional curvature satisfies Eq. (7.1.1) and so it is a symmetric Riemannian space due to Theorem 5.4.

Foot-notes:

[36)] The symbol (n) above Π is used here for distinction with projective connection parameters of Π^i_{jk} of Douglas [189].

[37] This can be directly obtained by transvecting Eq. (5.3) by $y_{;i}\dot{x}^h$ and using Eq. (4.9.15b), (4.9.21a) and $y_{;i}H^{i}_{jk} = 0$ [179].

CHAPTER 13

ON FINSLER SPACES WITH CONCIRCULAR TRANSFORMATIONS – II

§ 1. Introduction

Analogous to a symmetric Finsler space [108], we define a Ricci-symmetric Finsler space in terms of the Berwald's contracted curvature tensor (cf. Eq. (2.1) below). The discussion of concircular infinitesimal transformations is continued here for such spaces. The study of such transformations is concluded with the Finsler spaces *HR*-F_n of recurrent curvature (in the sense of Berwald) undergoing such transformations.

As shown in § 9 of Chapter 5, the curvature tensor along with its contracted form satisfy Eqs. (5.9.17), (5.9.18) and lead to Eq. (5.9.15a) on elimination of the terms $\rho\mu_k - \rho_k$ in a Finsler space with a c.i.t. These equations also hold good in a recurrent Finsler space (cf. Chapter 8, § 6) admitting a c.i.t. (called a *CHR*-F_n, $n > 2$, [135]). In addition, a covariant derivation of Eq. (5.9.15a), for Eqs. (5.9.1e), (8.1.1), (8.2.2a) and (5.9.15a) itself, also yields

$$H^{i}_{jkh} = \{2/(n-1)\}\, \delta^{i}_{[j}\, H_{k]h} \tag{1.1}$$

in a *CHR*-F_n, $n > 2$, which is same as Eqs. (8.6.2), (11.2.8) and (12.5.3). As mentioned in § 2 of Chapter 11, above relation causes symmetry of the contracted curvature tensor H_{kh} (cf. [135], Theorem 3.1) in a *CHR*-F_n, $n > 2$. This implies vanishing of the tensor H^{i}_{jki}:

$$H^{i}_{jki} = 0\,, \tag{1.2}$$

as well as vanishing of the projective curvature tensor W^{i}_{jkh} and its associated tensors W^{i}_{jk} and W^{i}_{j}.

§ 2. Ricci-symmetric F_n with a c.i.t.

Analogous to the symmetric Finsler space [108] we define a Ricci-symmetric Finsler space by

$$\mathfrak{B}_j\, H_{kh} = 0. \tag{2.1}$$

Successive transvections of this equation with directional coordinates, for Eqs. (4.6.7c) and (4.9.21), also yield Eqs. (7.2.3b) and (7.2.3c) in such a space. The directional derivation of Eq. (7.2.3b)), for the commutation formula (4.8.3), applied to the curvature vector H_k, and use of Eqs. (4.9.21b) and (2.1) there follows the identity:

$$H_i \, G^i_{jkh} = 0, \tag{2.2}$$

which is same as Eq. (7.2.7) holding in a symmetric space. On the other hand, the commutation formula (4.7.14), when applied to H_h, in view of Eqs. (4.9.21b) and (7.2.3b), yields

$$H_i \, H^i_{jkh} + H_{hi} \, H^i_{jk} = 0. \tag{2.3}$$

Its transvection with $\dot{x}^h$, for Eq. (4.9.15b), determines

$$(H_i + \dot{x}^h H_{hi}) \, H^i_{jk} = 0. \tag{2.4}$$

Also, application of commutation formula (4.7.14) to the scalar curvature H, for Eq. (7.2.3c), similarly gives

$$(\dot{\partial}_i H) \, H^i_{jk} = 0. \tag{2.5}$$

Transvecting Eqs. (2.4) and (2.5) by $\dot{x}^k$ and using Eq. (4.9.16), we get

$$(H_i + \dot{x}^h H_{hi}) \, H^i_j = 0 \quad (2.6); \quad \text{and} \quad (\dot{\partial}_i H) \, H^i_j = 0 \quad (2.7)$$

respectively in a Ricci-symmetric F_n. Also, the contracted forms of generalized Bianchi identity in Eq. (4.10.5) with respect to different pairs of indices: i, j and i, l, for Eqs. (4.9.20a), (4.10.3) and (2.1), reduce to

$$\mathfrak{B}_i \, H^i_{khm} + \{2H^l_{i[k} G^i_{h]lm} + H^l_{kh} G^i_{lmi}\} = 0, \tag{2.8}$$

and

$$H^m_{[jk} G^i_{h]mi} = 0. \tag{2.9}$$

With these characteristics of a Ricci-symmetric F_n we now assume the existence of a c.i.t. in the space. Forming covariant derivative of Eq. (5.9.18), putting from Eqs. (5.9.1e), (5.9.18) and (2.1), we obtain

$$\rho H_{kj} = (n-1)\{2\rho_{(j}\mu_{k)} + \rho(\mathfrak{B}_j\mu_k - \mu_j\mu_k) - \mathfrak{B}_j\rho_k\}. \qquad (2.10)$$

From Eqs. (5.9.2) and (5.9.11b) it is clear that the derivatives $\mathfrak{B}_j\mu_k$ and $\mathfrak{B}_j\rho_k$ are symmetric in the indices j, k. Thus, all the terms on the right side of Eq. (2.10) are symmetric implying the following result:

Theorem 2.1. The Ricci-like contracted curvature tensor H_{kj} is symmetric in a Ricci-symmetric Finsler space admitting a c.i.t.:

$$\text{(a)}\quad H_{jk} = H_{kj} \qquad \Rightarrow \qquad \text{(b)}\quad H^i_{jki} = 0. \qquad (2.11)$$

It may be noted that this result also holds even in a symmetric Finsler space with a c.i.t. (cf. Eq. (12.6.1)). For Eq. (4.9.21b), the identity (2.11a) is rewritten as

$$\dot{\partial}_k H_j = \dot{\partial}_j H_k, \qquad (2.11c)$$

that shows exactness of the differential equation $H_j d\dot{x}^j = 0$, having solution in the form:

$$H_j = \dot{\partial}_j J \qquad (2.12)$$

for some non-null scalar function J. Because of the homogeneous properties of H_j the function J is also positively homogeneous of degree two in its directional arguments. Hence, a transvection of Eq. (2.12) with $\dot{x}^j$ determines

$$J = \{(n-1)/2\}\, H, \qquad (2.13)$$

for Eq. (4.9.21c). Accordingly, Eq. (2.12) assumes the form

$$H_j = \{(n-1)/2\}\,\dot{\partial}_j H, \qquad (2.14)$$

with its directional derivative

$$H_{jk} = \{(n-1)/2\}\,\dot{\partial}_j\dot{\partial}_k H, \qquad (2.15)$$

due to Eq. (4.9.21b), in a Ricci-symmetric Finsler space admitting a c.i.t. In consequence of Eqs. (2.11) and the homogeneous properties of H_{jh} the projective curvature tensor, given by Eq. (5.2.15), assumes the form:

$$W^i_{jkh} = H^i_{jkh} - \{2/(n-1)\}\,\delta^i_{[j} H_{k]h}. \qquad (2.16)$$

Thus, the tensor M^{i}_{jkh}, defined by Eq. (12.2.3), coincides with the projecttive curvature tensor in the space under consideration. Hence, a transvection of above equation by v^h, for Eq. (5.9.15a), yields

$$W^{i}_{jkh}v^{h}=0 \tag{2.17}$$

in a Ricci-symmetric F_n, $n>2$, admitting a c.i.t. Successive transvections of Eq. (2.16) by $\dot{x}^{i}$'s, for Eqs. (4.9.15b), (4.9.16), (4.9.21), (5.2.16) and (5.2.17), also yield

$$W^{i}_{jk} = H^{i}_{jk} - \{2/(n-1)\,\delta^{i}_{[j}H_{k]}, \tag{2.18}$$

$$W^{i}_{j} = H^{i}_{j} - H\,\delta^{i}_{j} + \dot{x}^{i}\,H_{j}\,/\,(n-1). \tag{2.19}$$

Unlike a symmetric Finsler space with a c.i.t. becoming projectively flat (cf. Theorem 12.6.1), it follows from Eq. (2.16) that the projective curvature tensor need not be zero in a Ricci-symmetric Finsler space with a c.i.t. We, therefore, have the

Theorem 2.2. A Ricci-symmetric Finsler space with a c.i.t. need not be projectively flat.

In the following we observe that symmetry of the tensor H_{kh} plays a vital role in the geometry of a Ricci-symmetric Finsler space admitting a c.i.t. Forming covariant derivative of Eq. (5.9.15a) and putting from Eqs. (5.9.1e), (5.9.15a) itself and (2.1), we derive

$$v^{h}\mathfrak{B}_{l}\,H^{i}_{jkh} + \rho\left\{H^{i}_{jkl} - 2\delta^{i}_{[j}H_{k]l}\,/(n-1)\right\}=0. \tag{2.20}$$

Its skew-symmetric part with respect to the indices l, j, k, for Eqs. (4.10.1a) and (2.11), yields

$$v^{h}\mathfrak{B}_{[l}\,H^{i}_{jk]h}=0. \tag{2.21}$$

When contracted for indices i, l, and Eqs. (4.9.20a) and (2.1) used, above relation also implies

$$v^{h}\mathfrak{B}_{i}\,H^{i}_{jkh}=0. \tag{2.22}$$

On the other hand, the identity (2.4), for Eqs. (4.9.21a) and (2.11a), reduces to

$$H_{i}H^{i}_{jk}=0, \tag{2.23}$$

in a Ricci-symmetric Finsler space with a c.i.t. We note that the relations (2.5) and (2.23) also confirm a linear relation between the curvature vectors H_i and $\dot{\partial}_i H$ as expressed by Eq. (2.14) in such a space. A transvection of Eq. (2.18) by H_i, for Eq. (2.23), then yields

$$H_i W^i_{jk} = 0. \tag{2.24}$$

§ 3. *HR-F*

While deriving Eq. (10) Pandey [175] assumes equivalence of H/F^2 and the Riemannian curvature R of F_n, $n > 2$:

$$R = H / F^2, \tag{3.1}$$

and concludes the invariance of R implying vanishing nature of its covariant derivative. In fact, Eq. (3.1) holds only when F_n has the constant Riemannian curvature R [189], p.147. Thus, the Eq. (1.1) implies Eq. (11.2.10a) etc. (also cf. [135], Eq. (3.8), [119], Eq. (2.10b)):

$$H^i_{jk} = \{2/(n-1)\}\, \delta^i_{[j} H_{k]} = (2H/F^2)\, \delta^i_{[j} y_{k]} \tag{3.2}$$

against the Eq. (10) of [175]:

$$H^i_{jk} = 2R\delta^i_{[j} y_{k]} \tag{3.3}$$

as claimed by him. Hence, further analysis in [175] is erroneous and his Theorem 1 (denying the existence of a *CHR*-F_n, $n > 2$) is not correct. We, therefore, continue the discussion of existence of a c.i.t. in an *HR*-F_n, $n > 2$, in the following.

Equation (5.9.18) determines

$$\mu_k = \mathfrak{B}_k \ln \rho + H_{kh} v^h / (n-1)\,\rho \tag{3.4}$$

in any Finsler space admitting a c.i.t. Taking its covariant derivative and putting from Eqs. (5.9.1e), (8.2.2a) and (3.4) itself, we derive

$$\mathfrak{B}_j \mu_k = \mathfrak{B}_j \mathfrak{B}_k \ln \rho + (H_{ja} v^a)(H_{kh} v^h) / (n-1)^2 \rho^2$$
$$+ (\lambda_j H_{kh} v^h + \rho H_{kj}) / (n-1)\,\rho. \tag{3.5}$$

For Eqs. (5.9.2), (5.9.11b) and (1.2), the skew-symmetric part of Eq.

(3.5) with respect to the indices j, k yields

$$\lambda_{[j} H_{k]h} v^h = 0, \tag{3.6a}$$

$$\lambda_{[j} (\rho\, \mu_{k]} - \rho_{k]}) = 0, \tag{3.6b}$$

by Eq. (5.9.18) in a *CHR-F*$_n$, $n > 2$. Differentiating Eq. (3.6b) with respect to $\dot{x}^h$, transvecting the result so obtained by $\dot{x}^j$ and using Eqs. (5.9.10), (5.9.11a), (5.10.5), (8.2.4a), Note 5.9.2, and the homogeneous properties of μ, we get

$$\dot{\partial}_h \mu_k = 0 \quad (3.7a); \qquad \text{for} \quad \lambda \equiv \lambda_j \dot{x}^j \neq 0. \tag{3.7b}$$

Thus, μ_k becomes independent of the directional arguments in a *CHR-F*$_n$, $n > 2$, and so is the case in a flat Finsler space admitting a c.i.t. (cf. Eq. (12.3.1). Accordingly, Theorem 4.1 of [135] is amended as:

Theorem 3.1. In a *CHR-F*$_n$, $n > 2$, the vector μ_k too is a point function.

Applying the commutation formula vide Eq. (4.8.3) to the vector-field v^i, for Eqs. (5.9.1e), there follows the Eq. (5.9.19). The same, for Eqs. (5.9.11a) and (3.7a), reduces to

$$G^i_{jkh} v^h = 0 \tag{3.8}$$

in a *CHR-F*$_n$, $n > 2$, as well. Accordingly, the Lie derivative of G^i_{jk} given by Eq. (6.3.7), for Eqs. (4.10.15), (5.9.1e), (5.9.7) assumes the form

$$\pounds G^i_{jk} = \rho\, \delta^i_j \mu_k + \rho_j \delta^i_k + (\mathfrak{B}_j \mu_k + \mu_j \mu_k) v^i + v^h H^i_{hjk}, \tag{3.9}$$

together with

$$\pounds G^i_{ji} = \rho\, \mu_j + n\, \rho_j + (\mathfrak{B}_j \mu_i + \mu_j \mu_i) v^i, \tag{3.10}$$

by Eq. (1.2). A directional derivative of Eq. (3.9), for Eqs. (4.8.3), (5.9.10), (5.9.11a), (6.6.12), (3.7a) and Note 5.9.2, yields

$$\pounds G^i_{jkh} = -v^i \mu_a G^a_{jkh} + v^a \dot{\partial}_h H^i_{ajk}. \tag{3.11}$$

Contraction of Eq. (3.11) with respect to the indices i and h, for Eqs. (1.2), (3.8) and

$$\dot{\partial}_i H^{i}_{ajk} = \dot{\partial}_k H^{i}_{aji} = 0, \qquad \text{by Eqs. (4.9.15a),}$$

yields

$$£ G^{i}_{jki} = 0. \tag{3.12}$$

Therefore, for Eq. (6.2.9b), the Lie derivative of *normal projective connection* parameters given by Eq. (10.4.1) coincide with that of Berwald's connection parameters:

$$£ \overset{(n)}{\Pi}{}^{i}_{jk} = £ G^{i}_{jk}. \tag{3.13}$$

Same is the case in flat Finsler manifolds admitting a c.i.t. (cf. Eq. (12.3.10). Consequently, Theorem 12.3.1 also holds good in a $CHR\text{-}F_n$, $n > 2$:

Theorem 3.2. If an $HR\text{-}F_n$, $n > 2$, admits a projective motion defined by a c.i.t. then such a motion is necessarily of a restricted form:

$$£ G^{i}_{jk} = 2\delta^{i}_{(j} p_{k)}. \tag{3.14}$$

The hypothesis of the existence of a c.i.t. in an $HR\text{-}F_n$, $n > 2$, results in the relations connecting Berwald's tensor H^{i}_{jk} and the recurrence vector λ_k ([135], Eq. (3.13)):

$$H^{i}_{jk} = (2H/\lambda)\delta^{i}_{[j}\lambda_{k]}. \tag{3.15}$$

Consequently, its directional derivation, for Eqs. (4.9.15a), (6.1.3), (8.2.4a) and

$$\dot{\partial}_h \lambda = \lambda_h \quad \text{(3.16a)}; \qquad \lambda_h / \lambda = \dot{\partial}_h \ln \lambda, \quad \text{(3.16b)}$$

determines the curvature tensor:

$$H^{i}_{jkh} = (2/\lambda)\delta^{i}_{[j}\lambda_{k]}\{\dot{\partial}_h H - (H/\lambda)\lambda_h\} \tag{3.17a}$$

$$= 2\{\delta^{i}_{[j}\dot{\partial}_{k]} \ln \lambda\}\dot{\partial}_h H - H^{i}_{jk}\dot{\partial}_h \ln \lambda, \tag{3.17b}$$

by Eqs. (3.15) and (3.16). Also, for Eqs. (5.9.10), (5.9.11a) and (3.7a), it follows from Eqs. (5.9.17) and (5.9.18) that Berwald's curvature tensor H^{i}_{jkh} along with its contracted form H_{kh}, when transvected by v^h, loose their dependence upon the directional arguments in such space.

We check in the following if this independence can reduce the tensors H^{i}_{jkh} and H_{kh} merely as the point functions? Forming the covariant derivative of Eq. (5.9.18) and using Eqs. (8.2.2a), (5.9.1e) and (5.9.18) itself, we obtain

$$\rho H_{kj} = (n-1)\{2\rho_{(j}\,\mu_{k)} - \rho\mu_j\,\mu_k$$

$$+\rho\,(\mathfrak{B}_j - \lambda_j)\mu_k - (\mathfrak{B}_j - \lambda_j)\rho_k\}. \qquad (3.18)$$

Similarly, beginning with the covariant derivative of Eq. (5.9.17), we derive

$$\rho H^{i}_{jkl} = 2\,[(\mathfrak{B}_l - \lambda_l)\rho_{[j} - \{\rho_l + \rho\,(\mathfrak{B}_l - \lambda_l)\mu_{[j}\,\}]\,\delta^{i}_{k]}$$

$$-H^{i}_{jkh}\,v^h\,\mu_l\,. \qquad (3.19)$$

Thus, for $\mathfrak{B}_j\,\mu_k$ and $\mathfrak{B}_j\,\rho_k$ depending upon the directional arguments it appears that the tensors H_{kj} and H^{i}_{jkl} are not point functions. However, we continue the search in the following.

The Eq. (3.6b) implies a linear relation between the vector fields λ_k and $\rho\mu_k - \rho_k$:

$$T\lambda_k = \rho\mu_k - \rho_k \qquad (3.20a)$$

for some scalar point function T. But for Eq. (5.9.18), above relation can be rewritten as

$$T\lambda_k = (H_{kh}\,v^h)\,/\,(n-1). \qquad (3.20b)$$

Due to symmetry of H_{kj} in a $CHR\text{-}F_n$, $n > 2$, a transvection of Eq. (3.20b) by $\dot{x}^k$, for Eqs. (4.9.21a) and (3.7b), determines

$$T = (H_h\,v^h)\,/\,(n-1)\,\lambda. \qquad (3.21)$$

Hence, (3.20b) further reduces to

$$H_{kh}\,v^h = (\lambda_k\,/\,\lambda)\,H_h\,v^h = (\dot{\partial}_k \ln\lambda)\,H_h\,v^h, \qquad (3.20c)$$

by Eq. (3.16b). As a $CHR\text{-}F_n$, $n > 2$, is projectively flat ([135], Corollary 3.1), there holds the equation ([89], Eq. (3.7)):

$$\mathfrak{B}_j \dot{\partial}_k \ln \lambda = 0, \tag{3.22}$$

a covariant derivation of Eq. (3.20c) w.r.t. x^j, for Eqs. (8.2.2), (5.9.1e), (3.20c) itself and (3.22) determines

$$H_{kh} = (\dot{\partial}_k \ln \lambda)\, H_h. \tag{3.23a}$$

Its further transvection by $\dot{x}^h$, for Eqs. (4.9.21) and (3.16b), also determines

$$H_k = (n-1)\, H(\dot{\partial}_k \ln \lambda) = (n-1)\,(H/\lambda)\,\lambda_k. \tag{3.24}$$

Hence, Eq. (3.23a) reduces to

$$H_{kh} = (n-1)\, H(\dot{\partial}_k \ln \lambda)\,(\dot{\partial}_h \ln \lambda) = (n-1)\,(H/\lambda^2)\,\lambda_k \lambda_h, \tag{3.23b}$$

by Eq. (3.16a). Accordingly Eq. (1.1) reduces to

$$H^i_{jkh} = (2H/\lambda^2)\, \delta^i_{[j}\lambda_{k]}\lambda_h. \tag{3.25}$$

Also a directional derivative of Eq. (3.24), for Eqs. (4.9.21b), (3.16) and (3.23b), determines

$$\dot{\partial}_h H = 2H\, \dot{\partial}_h \ln \lambda = (2H/\lambda)\,\lambda_h, \tag{3.26}$$

which establishes an equivalence of Eqs. (3.17a) and (3.25). For Eq. (3.16a) we, therefore, have

$$\dot{\partial}_j (H/\lambda^2) = (\dot{\partial}_j H)/\lambda^2 - (2H/\lambda^3)\,\lambda_j = 0, \tag{3.27}$$

by Eq. (3.26), establishing the independence of H/λ^2 upon the directional arguments. Consequently, it follows from Eqs. (3.23b) and (3.25) that same is the case with the curvature tensors H_{kh} and H^i_{jkh} in a *CHR-*F_n, $n > 2$. Thus, we have established the:

Theorem 3.3. The curvature tensor of a *CHR-*F_n, $n > 2$, *is* a point function.

On the other hand, for Eqs. (3.16a) and (3.26), we note that

$$\dot{\partial}_j (H/\lambda) = (\dot{\partial}_j H)/\lambda - (H/\lambda^2)\,\lambda_j = (H/\lambda^2)\,\lambda_j$$

does not vanish in general. However, for Eqs. (8.2.4a) and (3.27), it is also seen as independent of the directional arguments. Hence, H / λ is not independent of $\dot{x}^i$'s. Therefore, it follows from Eq. (3.24) that same is the case with the curvature vector H_k . Thereby, it is also evident from Eqs. (4.9.21c) and (3.2) that the tensors H^i_{jk} as well as

$$H^i_j \equiv H^i_{jk}\,\dot{x}^k = H\,\delta^i_j - \dot{x}^i H_j /(n-1) \qquad (3.28)$$

are not independent of $\dot{x}^i$'s.

Forming covariant derivative of Eq. (1.1) with respect to x^l, taking skew-symmetric part of the derived equation with respect to the indices l, j, k and using Eq. (4.10.5), we derive

$$H^a_{[kj} G^i_{l]ha} = \{2/(n-1)\,\lambda_{[j}\delta^i_k H_{l]h}, \qquad (3.29)$$

in a $CHR\text{-}F_n$, $n > 2$. The left side of Eq. (3.29) vanishes identically for the tensor H^a_{kj} given by Eq. (3.2) and the symmetry of G^i_{lha} in its lower indices, thus, reducing Eq. (3.29) to

$$\lambda_{[j}\delta^i_k H_{l]h} = 0 \quad (3.30) \qquad \Rightarrow \qquad 2\lambda_{[k} H_{l]h} = 0, \qquad (3.31)$$

on contraction for i and j for $n > 2$. Thus, the relations (3.23b) and (3.24) are also deducible from Eq. (3.31) on its successive transvections by the directional arguments in view of Eqs. (4.9.21a), (1.2) and (3.7b).

CHAPTER 14

AN AXIOMATIC APPROACH TO TENSORS

The present chapter offers an exposition of the axiomatic definition of tensors and their further developments from this very standpoint. Various types of tensors and their examples have been included.

§ 1. Introduction

The classical definition of a tensor was introduced by Italian geometer Giovanni Ricci in 1987. Thereafter, tensors have been extensively used as tools in the study of Differential Geometry, Theory of Relativity, Physics, Mechanics, Engineering and many other allied fields. However, the contemporary treatment of tensors is relatively new and its necessity arose mainly due to the following drawbacks in the classical definition:

(i) The components of tensors are not well defined;

(ii) The transformation rules for the tensors relate the components of tensors but not the tensors themselves;

(iii) In the transformation rules there occur the derivatives $\partial \bar{x}^i / \partial x^j$ which become meaningless if the space under consideration is non-differentiable;

(iv) The sum of two tensors of different types appears to exist but, indeed, it does not.

Owing to these drawbacks the celebrated French geometer Elié Cartan introduced a new definition of a tensor in the early years of the last century. The new definition of tensors is free from above drawbacks. Elié Cartan's approach to tensors, which forms the subject matter of the present chapter, not only gives a rigorous foundation of tensors but also stimulates a feeling for the beauty and interest of rigor.

§ 2. Vector space. Contravariant vectors

Let F be a field whose elements are scalars (usually, real numbers) and V a non-empty set of elements called vectors. Let a binary operation of vector addition and an operation of scalar multiplication satisfying the following axioms be defined on V:

2.1. $(V, +)$ is an abelian group, i.e.

i) $\forall\ \boldsymbol{\lambda}, \boldsymbol{\mu}\ \varepsilon\ V \Rightarrow \boldsymbol{\lambda} + \boldsymbol{\mu}\ \varepsilon\ V$ (binary property),

ii) $\forall\ \boldsymbol{\lambda}, \boldsymbol{\mu}, \boldsymbol{\upsilon}\ \varepsilon\ V \Rightarrow \boldsymbol{\lambda} + (\boldsymbol{\mu} + \boldsymbol{\upsilon}) = (\boldsymbol{\lambda} + \boldsymbol{\mu}) + \boldsymbol{\upsilon}$ (associative law),

iii) $\forall\ \boldsymbol{\lambda}\ \varepsilon\ V\ \exists\ \mathbf{0}\ \varepsilon\ V \Rightarrow \boldsymbol{\lambda} + \mathbf{0} = \mathbf{0} + \boldsymbol{\lambda}$ (identity law),

iv) $\forall\ \boldsymbol{\lambda}\ \varepsilon\ V\ \exists\ -\boldsymbol{\lambda}\ \varepsilon\ V \Rightarrow \boldsymbol{\lambda} + (-\boldsymbol{\lambda}) = (-\boldsymbol{\lambda}) + \boldsymbol{\lambda} = \mathbf{0}$ (inverse law),

v) $\forall\ \boldsymbol{\lambda}, \boldsymbol{\mu}\ \varepsilon\ V \Rightarrow \boldsymbol{\lambda} + \boldsymbol{\mu} = \boldsymbol{\mu} + \boldsymbol{\lambda}$ (commutative law);

2.2. $\forall\ a\ \varepsilon\ F$ and $\boldsymbol{\lambda}\ \varepsilon\ V$ there corresponds a unique vector $a\boldsymbol{\lambda}$, called the scalar multiple of $\boldsymbol{\lambda}$ by a such that the scalar multiplication is associative, i.e.

i) $a, b\ \varepsilon\ F, \boldsymbol{\lambda}\ \varepsilon\ V \quad \Rightarrow \quad a\,(b\,\boldsymbol{\lambda}) = (ab)\,\boldsymbol{\lambda},$ ii) $1\boldsymbol{\lambda} = \boldsymbol{\lambda};$

2.3. i) Multiplication by a scalar is distributive over vector addition:

$$a\,(\boldsymbol{\lambda} + \boldsymbol{\mu}) = a\,\boldsymbol{\lambda} + a\,\boldsymbol{\mu}, \quad a\ \varepsilon\ F \text{ and } \boldsymbol{\lambda}, \boldsymbol{\mu}\ \varepsilon\ V;$$

ii) Multiplication by a vector is distributive over scalar addition:

$$(a + b)\,\boldsymbol{\lambda} = a\,\boldsymbol{\lambda} + b\,\boldsymbol{\lambda}, \ a, b\ \varepsilon\ F \text{ and } \boldsymbol{\lambda}\ \varepsilon\ V,$$

where (+) in the left side is addition in F but in the right side it refers to addition in V.

Definition 2.1. The structure $(V, +, \cdot)$ containing the set of vectors V and two operations, namely vector addition and scalar multiplication satisfying above axioms 2.1 to 2.3 is called a *vector space* over the field F.

For convenience, this vector space will be denoted by V^1.

Example 2.1. The set of all ordered n-tuples $\mathbf{a} = (a^1, a^2, \ldots, a^n)$, $\mathbf{b} = (b^1, b^2, \ldots, b^n)$, etc. of elements of a field F is a vector space with respect to addition defined by

$$\mathbf{a}+\mathbf{b} = (a^1+b^1, a^2+b^2, \dots, a^n+b^n),$$

and scalar multiplication defined by

$$c\mathbf{a} = (ca^1, ca^2, \dots, ca^n),\ c \,\varepsilon\, F.$$

Definition 2.2. A set S of linearly independent vectors of V, such that every vector in V is a linear combination of elements of S, is called a *basis* of the vector space V^1; and the number of elements of S is called the *dimension* of the space.

Let n be the dimension of the vector space and a basis of the space be given by the set of linearly independent vectors $\mathbf{e}_1, \mathbf{e}_2, \dots, \mathbf{e}_n$. For brevity, this set may be denoted by $\{\mathbf{e}_i\}$ only, where the index i takes the positive integral values from 1 to n. (Throughout our discussion the dimension of the vector space V^1 is assumed n.) If the set $\{\bar{\mathbf{e}}_i\}$ forms another basis of the space it then follows from Definition 2.2 that each vector of this set is linearly dependent upon the vectors of the set $\{\mathbf{e}_i\}$ and vice-versa. Consequently, there hold the relations

$$\bar{\mathbf{e}}_i = p_i^j\,\mathbf{e}_j \quad (2.1\text{a}); \qquad \mathbf{e}_i = q_i^j\,\bar{\mathbf{e}}_j \quad (2.1\text{b});$$

and the matrices of coefficients $((p_i^j))$ and $((q_i^j))$ are inverse to each other.

Any vector $\boldsymbol{\lambda}$ of the space V^1, being linearly dependent upon the basis vectors, can be written as a linear combination of the latter ones:

$$\boldsymbol{\lambda} = \lambda^i\mathbf{e}_i, \tag{2.2}$$

where the coefficients λ^i are scalars belonging to the field F and are called the components of the vector $\boldsymbol{\lambda}$ with regard to the basis $\{\mathbf{e}_i\}$. It may be easily seen from the equations (2.1) and (2.2) that the components λ^i and $\bar{\lambda}^i$ of the vector $\boldsymbol{\lambda}$ with regard to the bases $\{\mathbf{e}_i\}$ and $\{\bar{\mathbf{e}}_i\}$ obey the following transformation laws:

$$\bar{\lambda}^i = q_j^i\lambda^j \quad (2.3\text{a}); \qquad \lambda^i = p_j^i\bar{\lambda}^j. \quad (2.3\text{b})$$

Definition 2.3. Elements of the vector space discussed above are called the *contravariant vectors*.

§ 3. Dual space. Covariant vectors

It will be shown in this section that, associated to a vector space V^1, there corresponds another vector space (to be denoted by V_1). The latter shall be called a *dual* (*vector*) *space* of the former one. Elements of the dual space will be called *covariant vectors*. To begin with this, we shall first define a linear scalar function on the vector space V^1.

Definition 3.1. A linear mapping $\boldsymbol{\alpha}$: $V^1 \to F$ of the vector space V^1 over the field F defines a *linear scalar function* if it satisfies

$$\boldsymbol{\alpha}(\boldsymbol{\lambda} + \boldsymbol{\mu}) = \boldsymbol{\alpha}(\boldsymbol{\lambda}) + \boldsymbol{\alpha}(\boldsymbol{\mu}), \qquad \boldsymbol{\lambda}, \boldsymbol{\mu} \in V^1; \tag{3.1a}$$

$$\boldsymbol{\alpha}(a\,\boldsymbol{\lambda}) = a\,\boldsymbol{\alpha}(\boldsymbol{\lambda}), \quad a \in F, \ \boldsymbol{\lambda} \in V^1. \tag{3.1b}$$

Let us now consider a system V_1 consisting of the set of all such linear scalar functions over V^1, a binary operation of vector addition as defined above, vide Eqn. (3.1a), and another operation of scalar multiplication satisfying

$$(a\,\boldsymbol{\alpha})(\boldsymbol{\lambda}) = a\,\{\boldsymbol{\alpha}(\boldsymbol{\lambda})\}, \ a \in F, \ \boldsymbol{\lambda} \in V^1, \ \boldsymbol{\alpha} \in V_1. \tag{3.2}$$

It is easy to verify that V_1 satisfies all the postulates of a vector space. This vector space is called *dual space* of the vector space V^1.

Definition 3.2. The set of linearly independent vectors $\mathbf{e}^i$ satisfying

$$\mathbf{e}^i(\mathbf{e}_j) = \delta^i_j = 1 \quad (\text{if } i = j), \quad 0 \quad (\text{if } i \neq j), \tag{3.3}$$

form a basis of the dual space V_1, where $\mathbf{e}_j$ are pre-supposed to form a basis of the vector space V^1.

A basis of the dual space is called a *dual basis* of the vector space V^1. It may be seen that two bases $\{\mathbf{e}^i\}$ and $\{\bar{\mathbf{e}}^i\}$ of the dual space V_1 satisfy

$$\bar{\mathbf{e}}^i = q^i_j\,\mathbf{e}^j \quad (3.4a); \qquad \mathbf{e}^i = p^i_j\,\bar{\mathbf{e}}^j. \quad (3.4b)$$

Analogous to Eqn. (2.2), a vector $\boldsymbol{\alpha} \in V_1$ can be linearly expressed in terms of the basis vectors:

$$\boldsymbol{\alpha} = \alpha_i \, \mathbf{e}^i. \tag{3.5}$$

The coefficients α_i are called components of the covariant vector $\boldsymbol{\alpha}$. Let $\overline{\alpha}_i$ be the components of the vector $\boldsymbol{\alpha}$ with regard to the basis $\{\overline{\mathbf{e}}^i\}$. Then these are connected to α_i by following transformation rules

$$\overline{\alpha}_i = p_i^j \, \alpha_j \quad \text{(3.6a)}; \qquad \alpha_i = q_i^j \, \overline{\alpha}_j. \quad \text{(3.6b)}$$

§ 4. Tensor product of vector spaces. Tensors

Let V and W be two vector spaces defined over a field F with their dual spaces V^* and W^* respectively. The Cartesian product $V^* \times W^*$ is the set of all ordered pairs $(\boldsymbol{\lambda}^*, \boldsymbol{\alpha}^*)$, $\boldsymbol{\lambda}^* \, \varepsilon \, V^*$, $\boldsymbol{\alpha}^* \, \varepsilon \, W^*$. We shall now define a bilinear scalar function over this product.

Definition 4.1. A bilinear mapping $\mathbf{T}$: $V^* \times W^* \rightarrow F$ defines a *bilinear scalar function* over $V^* \times W^*$ provided $\mathbf{T}(\boldsymbol{\lambda}^*, \boldsymbol{\alpha}^*)$ is a scalar function satisfying

$$\left.\begin{aligned} \mathbf{T}(\boldsymbol{\lambda}^* + \boldsymbol{\mu}^*, \boldsymbol{\alpha}^*) &= \mathbf{T}(\boldsymbol{\lambda}^*, \boldsymbol{\alpha}^*) + \mathbf{T}(\boldsymbol{\mu}^*, \boldsymbol{\alpha}^*), \\ \mathbf{T}(\boldsymbol{\lambda}^*, \boldsymbol{\alpha}^* + \boldsymbol{\beta}^*) &= \mathbf{T}(\boldsymbol{\lambda}^*, \boldsymbol{\alpha}^*) + \mathbf{T}(\boldsymbol{\lambda}^*, \boldsymbol{\beta}^*), \\ \mathbf{T}(a\,\boldsymbol{\lambda}^*, \boldsymbol{\alpha}^*) &= \mathbf{T}(\boldsymbol{\lambda}^*, a\,\boldsymbol{\alpha}^*) = a\,\mathbf{T}(\boldsymbol{\lambda}^*, \boldsymbol{\alpha}^*), \end{aligned}\right\} \tag{4.1}$$

where $\boldsymbol{\lambda}^*, \boldsymbol{\mu}^* \, \varepsilon \, V^*$, $\boldsymbol{\alpha}^*, \boldsymbol{\beta}^* \, \varepsilon \, W^*$ and $a \, \varepsilon \, F$.

The set of all such real-valued bilinear scalar functions form a vector space over the same field F under the operations of vector addition and scalar multiplication defined by

$$\left.\begin{aligned} (\mathbf{T} + \mathbf{S})(\boldsymbol{\lambda}^*, \boldsymbol{\alpha}^*) &= \mathbf{T}(\boldsymbol{\lambda}^*, \boldsymbol{\alpha}^*) + \mathbf{S}(\boldsymbol{\lambda}^*, \boldsymbol{\alpha}^*), \\ (a\,\mathbf{T})(\boldsymbol{\lambda}^*, \boldsymbol{\alpha}^*) &= a\,\{\mathbf{T}(\boldsymbol{\lambda}^*, \boldsymbol{\alpha}^*)\}. \end{aligned}\right\} \tag{4.2}$$

Definition 4.2. The vector space obtained above is called the *tensor product* of the vector spaces V and W, and is denoted by $V \otimes W$. Elements of this space (which are also vectors with regard to the bases of $V \otimes W$) are called *tensors of second order* with regard to the bases of V and W.

Note 4.1. It may be noted that to define the tensor product of spaces V and W, the bilinear functions are defined over the Cartesian product of their dual spaces.

In particular, we now begin with a vector space V^1 and its dual V_1. We consider a system consisting of the set of all bilinear scalar functions over $V_1 \times V_1$, a binary operation of vector addition and another operation of scalar multiplication satisfying

$$(\mathbf{T} + \mathbf{S})(\boldsymbol{\alpha}, \boldsymbol{\beta}) = \mathbf{T}(\boldsymbol{\alpha}, \boldsymbol{\beta}) + \mathbf{S}(\boldsymbol{\alpha}, \boldsymbol{\beta}),\ (a\,\mathbf{T})(\boldsymbol{\alpha}, \boldsymbol{\beta}) = a\,\{\mathbf{T}(\boldsymbol{\alpha}, \boldsymbol{\beta})\}. \quad (4.3)$$

where $\boldsymbol{\alpha}, \boldsymbol{\beta}\ \varepsilon\ V_1$, $a\ \varepsilon$ F and $\mathbf{T}$, $\mathbf{S}$ are two such functions. It may be shown that this system is a vector space. As per Defn. 4.2, it will be called the tensor product of the vector space V^1 with itself and shall be denoted by $V^1 \otimes V^1$, or more conveniently, by V^2.

Let $\{\mathbf{e}^i\}$ be a basis of dual space V_1 and $\mathbf{e}_{kh}$ the bilinear functions mapping $(\mathbf{e}^i, \mathbf{e}^j)\ \varepsilon\ V_1 \times V_1$ into $\delta^i_k\ \delta^j_h\ \varepsilon\ F$:

$$\mathbf{e}_{k\,h}(\mathbf{e}^i, \mathbf{e}^j) = \delta^i_k\ \delta^j_h. \quad (4.4)$$

Hence, under this mapping $(\boldsymbol{\alpha}, \boldsymbol{\beta})\ \varepsilon\ V_1 \times V_1$ maps into $\alpha_k\ \beta_h$:

$$\mathbf{e}_{k\,h}(\boldsymbol{\alpha}, \boldsymbol{\beta}) = \mathbf{e}_{k\,h}(\alpha_i\,\mathbf{e}^i, \beta_j\,\mathbf{e}^j) = \alpha_i\ \beta_j\ \delta^i_k\ \delta^j_h = \alpha_k\ \beta_h, \quad (4.5)$$

where α_k (respectively β_h) are the components of $\boldsymbol{\alpha}$ (resp. $\boldsymbol{\beta}$) with regard to the basis $\{\mathbf{e}^i\}$ of V_1. Also, it may be shown that n^2 bilinear functions $\mathbf{e}_{k\,h}$, as considered above, are linearly independent to each other. As such, they can constitute a basis of the space V^2. Any vector $\mathbf{T}\ \varepsilon\ V^2$ becomes dependent upon these basis vectors:

$$\mathbf{T} = \mathrm{T}^{kh}\,\mathbf{e}_{k\,h}. \quad (4.6)$$

The coefficients T^{kh} are the components of the vector $\mathbf{T}\ \varepsilon\ V^2$ with regard to the basis $\{\mathbf{e}_{k\,h}\}$. When compared to the basis $\{\mathbf{e}_k\}$ of V^1 the vector $\mathbf{T}\ \varepsilon$ V^2 is regarded as a *contravariant tensor* of order two. If the dimension of V^1 is n that of V^2 is then n^2. So, there are n^2 components of the tensor $\mathbf{T}$.

Given two vectors $\boldsymbol{\lambda}, \boldsymbol{\mu}\ \varepsilon\ V^1$ their tensor product $\boldsymbol{\lambda} \otimes \boldsymbol{\mu}$ is a bilinear mapping taking the pair $(\boldsymbol{\lambda}, \boldsymbol{\mu})$ to a vector of V^2. The components of this vector of V^2 relative to the basis $\{\mathbf{e}_{i\,j}\}$ are $\lambda^i\ \mu^j$ if λ^i (respectively μ^j)

are components of the vector $\boldsymbol{\lambda}$ (resp. $\boldsymbol{\mu}$) relative to the basis $\{\mathbf{e}_i\}$ of V^1. Thus, we may write

$$\boldsymbol{\lambda} \otimes \boldsymbol{\mu} = \lambda^i \mu^j \mathbf{e}_{ij}. \qquad (4.7)$$

It may be easily seen that the tensor product of the vectors $\mathbf{e}_i$ and $\mathbf{e}_j$, which form a basis of V^1, is the vector $\mathbf{e}_{ij}$ of V^2:

$$\mathbf{e}_i \otimes \mathbf{e}_j = \mathbf{e}_{ij}. \qquad (4.8)$$

It can be also shown that the components T^{ij} and $\overline{T}^{ij}$ of a vector $\mathbf{T}$ ε V^2 with regard to the bases $\{\mathbf{e}_{ij}\}$ and $\{\overline{\mathbf{e}}_i\}$respectively are connected by

$$\overline{T}^{ij} = q^i_k\, q^j_h\, T^{kh} \qquad (4.9a); \quad T^{ij} = p^i_k\, p^j_h\, \overline{T}^{kh}. \qquad (4.9b)$$

§ 5. Various types of tensors

In the preceding section, we have defined a contravariant tensor of second order as a vector of the space V^2, where V^2 is constituted by the set of bilinear scalar functions defined over the Cartesian product of V_1 with itself. Similarly, we may consider the Cartesian product of the vector space V^1 with itself and can define the tensor product $V_1 \otimes V_1$ (more conveniently denoted by V_2). This system will consist of the set of all bilinear scalar functions $\mathbf{X}$, $\mathbf{Y}$, ... defined over $V^1 \times V^1$, a binary operation of vector addition and an operation of scalar multiplication satisfying

$$(\mathbf{X} + \mathbf{Y})(\boldsymbol{\lambda}, \boldsymbol{\mu}) = \mathbf{X}(\boldsymbol{\lambda}, \boldsymbol{\mu}) + \mathbf{Y}(\boldsymbol{\lambda}, \boldsymbol{\mu}), \qquad (5.1a)$$

$$(a\,\mathbf{X})(\boldsymbol{\lambda}, \boldsymbol{\mu}) = a\,\{\mathbf{X}(\boldsymbol{\lambda}, \boldsymbol{\mu})\}, \qquad (5.1b)$$

where $\boldsymbol{\lambda}$, $\boldsymbol{\mu}$ ε V^1, a ε F. We now define a bilinear function $\mathbf{e}^{kh}$ over $V^1 \times V^1$:

$$\mathbf{e}^{kh}(\mathbf{e}_i, \mathbf{e}_j) = \delta^k_i\, \delta^h_j. \qquad (5.2)$$

It may be shown that $\mathbf{e}^{kh}$ maps $(\boldsymbol{\lambda}, \boldsymbol{\mu})$ as follows:

$$\mathbf{e}^{kh}(\boldsymbol{\lambda}, \boldsymbol{\mu}) = \lambda^k \mu^h, \qquad (5.3)$$

where λ^k (respectively μ^h) are the components of the vector $\boldsymbol{\lambda}$ (resp. $\boldsymbol{\mu}$). Further, it may be also shown that the vectors $\mathbf{e}^{kh}$ are linearly independent with respect to each other. Hence, they form a basis of the space V_2.

Now we can define the tensor product $\boldsymbol{\alpha} \otimes \boldsymbol{\beta}$ of two vectors $\boldsymbol{\alpha}, \boldsymbol{\beta} \in V_1$ as a bilinear mapping taking the pair $(\boldsymbol{\alpha}, \boldsymbol{\beta})$ to a vector of V_2 with components $\alpha_i \beta_j$ with regard to the basis $\{\mathbf{e}^{ij}\}$ of V_2:

$$\boldsymbol{\alpha} \otimes \boldsymbol{\beta} = \alpha_i \beta_j \mathbf{e}^{ij}, \tag{5.4}$$

where α_i (respectively β_j) are components of the vector $\boldsymbol{\alpha}$ (resp. $\boldsymbol{\beta}$). It may be also shown that

$$\mathbf{e}^i \otimes \mathbf{e}^j = \mathbf{e}^{ij}. \tag{5.5}$$

Definition 5.1. Elements of the vector space V_2 (which are vectors with respect to the basis $\{\mathbf{e}^{ij}\}$ of V_2) are called the *covariant tensors of order two* with regard to the basis $\{\mathbf{e}^i\}$ of V_1 .

Let a vector $\mathbf{X} \in V_2$ have components X_{ij} (respectively $\overline{X}_{ij}$) with respect to the basis $\{\mathbf{e}^{ij}\}$ (resp. $\{\overline{\mathbf{e}}_i\}$). These components are connected by

$$\overline{X}_{ij} = p_i^k \, p_j^h \, X_{kh} \quad (5.6a); \quad X_{ij} = q_i^k \, q_j^h \, \overline{X}_{kh}. \quad (5.6b)$$

We have discussed, so far, two different types of second order tensors: contravariant and covariant tensors. Yet there is another kind of second order tensor, called a tensor of mixed type involving both the *contra* and *co*-nature. For such type of tensors we begin with a vector space V^1 and its dual V_1, and construct their Cartesian product $V_1 \times V^1$. As in the previous section, we then construct their tensor product $V^1 \otimes V_1$ (briefly denoted by V_1^1). It consists of the set of all bilinear scalar functions $\mathbf{A}, \mathbf{B}, \ldots$ defined over $V_1 \times V^1$, a binary operation of vector addition and an operation of scalar multiplication satisfying

$$(\mathbf{A} + \mathbf{B})(\boldsymbol{\alpha}, \boldsymbol{\lambda}) = \mathbf{A}(\boldsymbol{\alpha}, \boldsymbol{\lambda}) + \mathbf{B}(\boldsymbol{\alpha}, \boldsymbol{\lambda}), \tag{5.7a}$$

$$(a\,\mathbf{A})(\boldsymbol{\alpha}, \boldsymbol{\lambda}) = a\,\{\mathbf{A}(\boldsymbol{\alpha}, \boldsymbol{\lambda})\}, \tag{5.7b}$$

where $\boldsymbol{\alpha} \in V_1$, $\boldsymbol{\lambda} \in V^1$ and $a \in F$. Defining a bilinear scalar function $\mathbf{e}^i_j$ over $V_1 \times V^1$:

$$\mathbf{e}^i_j\,(\mathbf{e}^h, \mathbf{e}_k) = \delta^i_k\, \delta^h_j. \tag{5.8}$$

it may be seen that $\mathbf{e}^i_j$ maps $(\boldsymbol{\alpha}, \boldsymbol{\lambda}) \in V_1 \times V^1$ into $\alpha_j \lambda^i \in F$:

$$\mathbf{e}^{i}_{j}\,(\boldsymbol{\alpha}, \boldsymbol{\lambda}) = \alpha_j\, \lambda^i, \qquad \boldsymbol{\alpha} \,\varepsilon\, V_1, \;\; \boldsymbol{\lambda} \,\varepsilon\, V^1. \tag{5.9}$$

Further, the vectors $\mathbf{e}^{i}_{j}$ being linearly independent with respect to each other, form a basis of the space $V^{\,1}_{\,1}$.

Definition 5.2. Elements of the space $V^{\,1}_{\,1}$ (which are vectors with respect to the basis $\{\mathbf{e}^{i}_{j}\}$ of $V^{\,1}_{\,1}$) are called the *tensors of type* (1, 1) or *mixed tensors of second order* with regard to the bases $\{\mathbf{e}_i\}$ of V^1 and $\{\mathbf{e}^i\}$ of V_1.

Example 5.1. Kronecker deltas are components of a mixed tensor of second order.

The formalism of mixed tensors discussed above gives rise to a mixed tensor of arbitrary rank. In the following, we discuss the tensors of type (r, s) possessing the contravariant nature of order r and covariant nature of order s (both r and s being positive integers).

The tensor product $V^{\,1} \otimes V^{\,1} \otimes \ldots \otimes V^{\,1} \otimes V_1 \otimes V_1 \otimes \ldots \otimes V_1$ (where V^1 and V_1 repeat r and s times respectively), or more conveniently denoted by $V^{\,r}_{\,s}$, can be defined by a system consisting of the set of all $(r + s)$ - linear scalar functions defined over

$$V_1 \times V_1 \times \ldots \times V_1 \times V^1 \times V^1 \times \ldots \times V^1$$

(V_1 repeating r times while V^1 repeating s times),

a binary operation of vector addition and another operation of scalar multiplication satisfying

$$(\mathbf{P} + \mathbf{Q})\,(\boldsymbol{\alpha}_1, \boldsymbol{\alpha}_2, \ldots, \boldsymbol{\alpha}_r, \boldsymbol{\lambda}^1, \boldsymbol{\lambda}^2, \ldots, \boldsymbol{\lambda}^s) = \mathbf{P}\,(\boldsymbol{\alpha}_1, \boldsymbol{\alpha}_2, \ldots, \boldsymbol{\alpha}_r, \boldsymbol{\lambda}^1, \boldsymbol{\lambda}^2, \ldots, \boldsymbol{\lambda}^s)$$

$$+\, \mathbf{Q}\,(\boldsymbol{\alpha}_1, \boldsymbol{\alpha}_2, \ldots, \boldsymbol{\alpha}_r, \boldsymbol{\lambda}^1, \boldsymbol{\lambda}^2, \ldots, \boldsymbol{\lambda}^s), \tag{5.10a}$$

$$(a\,\mathbf{P})\,(\boldsymbol{\alpha}_1, \boldsymbol{\alpha}_2, \ldots, \boldsymbol{\alpha}_r, \boldsymbol{\lambda}^1, \boldsymbol{\lambda}^2, \ldots, \boldsymbol{\lambda}^s)$$

$$= a\,\{\mathbf{P}\,(\boldsymbol{\alpha}_1, \boldsymbol{\alpha}_2, \ldots, \boldsymbol{\alpha}_r, \boldsymbol{\lambda}^1, \boldsymbol{\lambda}^2, \ldots, \boldsymbol{\lambda}^s)\}, \tag{5.10b}$$

for $\boldsymbol{\alpha}_1, \boldsymbol{\alpha}_2, \ldots, \boldsymbol{\alpha}_r \; \varepsilon \; V_1, \boldsymbol{\lambda}^1, \boldsymbol{\lambda}^2, \ldots, \boldsymbol{\lambda}^s \; \varepsilon V^1$ and $a \; \varepsilon \; F$. It can thus be shown that V^r_s is a vector space and the functions $\mathbf{e}^{j_1 j_2 \ldots j_s}_{i_1 i_2 \ldots i_r}$ constitute a set of linearly independent n^{r+s} vectors - hence forming a basis of the vector space V^r_s.

Definition 5.3. Elements of the space V^r_s, which are obviously vectors with respect to the basis $\{\mathbf{e}^{j_1 j_2 \ldots j_s}_{i_1 i_2 \ldots i_r}\}$, are called the *mixed tensors of type* (r, s) when regarded to the bases $\{\mathbf{e}_i\}$ of V^1 and $\{\mathbf{e}^j\}$ of V_1 respectively.

Example 5.2. Riemann-Christoffel symbols R^i_{jkh} constitute the components of a mixed tensor of type (1, 3). This tensor is called the *curvature tensor* in Riemannian geometry.

CHAPTER 15

PHYSICAL FIELD THEORIES

§ 1. Introduction

In physical theories, fields are characterized not only by their invariance properties under transformations of the points x^i of the background space-time manifold (Lorentz invariance, space-time parity, etc.) but also by "internal symmetries" (isotropic invariance, SU(3)[38)] symmetry, charge parity, etc.), which are independent of x^i. The internal symmetry is often associated with the transformations of the points of an "internal space", the symmetry of which is reflected in the behaviour of physical fields. Comparison of this method with the theory of Finsler geometry suggests that some of the important classes of internal symmetries of physical fields may be conceived as manifestations of a Finslerian structure inherent in real space-time.

These facts lead to the hypothesis that the internal symmetries of physical fields reflect the symmetry properties of the indicatrix of a Finslerian structure assumed to be possessed by real space-time. Accordingly, the indicatrix plays the role of an internal space, whereas its points serve as the internal coordinates and the group of motions in the indicatrix play the role of a group of internal symmetries.

Following Drechsler and Mayer [40], the transformations $u^a = u^a(\tilde{u})$ may be regarded as global transformations and their x-dependent generalizations $u^a = u^a(x, \tilde{u})$ as local gauge transformations or merely gauge transformations. The gauge fields are constructible from the projection factors of the indicatrix of a Finsler space. The gauge-field strength tensors are expressible in terms of the Finslerian curvature tensors. The gauge-covariant differentiation operator can be readily extended to the case of spinor fields.

In physical theories gauge transformations are usually linear:

$$u^a = u^a_b(x)\,\tilde{u}^b.$$

An important example of such a theory is the Yang-Mills theory, which when restricted to SU(2) gauge transformations, describe the isotropic invariance of strongly interacting fields in terms of gauge fields, or weak interactions in the context of the standard model of unification of electromagnetic and weak interactions. These ideas readily account for the geometrical meaning of isotropic symmetry as a manifestation of the Finslerian structure of space-time with the highest symmetry indicatrix. It may be recalled that isotropic symmetry manifests itself as a symmetry of physical fields under Euclidean rotations in a three-dimensional internal space called the *isotropic space*. In spite of the fundamental role of isotropic invariance in the theory of physical fields, Heisenberg treated this invariance abstractly as a symmetry relative to Euclidean rotations in an imaginary internal space. The indicatrix of a four-dimensional Finsler space is a three-dimensional space, so that in the most symmetrical case, when the curvature tensors of the indicatrix vanish, the indicatrix represents a three-dimensional Euclidean space, which may be identified with isotropic space. A striking example of a metric function of such a Finsler space is the Berwald-Moór metric function. Also, the conformally flat 1-form Berwald-Moór Finsler space is not flat yet admits gauge fields independent of u^a. So, it reflects both isotropic invariance and some gravitational effects. The notion of embedding isotropic space into the tangent space of the space-time manifold is an important feature of Finslerian approach. The embedding is given by the parametrical representation of the indicatrix in the case of the Berwald-Moór metric function. Significantly, this relationship opens up the possibility of unifying the gravitational and Yang-Mills SU(2) gauge fields in terms of space-time geometric objects. The classical Yang-Mills gauge field equations are constructed on the basis of a Lagrangian, which is quadratic in the gauge-field strength tensor. It is shown by Meetz [86] that the Lagrangian for describing interactions of soft π-messons can be obtained by assuming its invariance under the group of motions of a three-dimensional Riemannian isotropic space of constant positive curvature. Meetz describes nucleons by functions of the form $\varphi^\beta \{x, p^a(x)\}$, which depend on the fields $p^a(x)$ of π-messons, φ^β is four component space-time spinor. The Finslerian techniques enable one to formulate gauge-invariant equations of physical fields for various internal symmetries in an arbitrary curved isotropic space.

Each component of Dirac spinor satisfies the Klein-Gordon equation.

§ 2. Finslerian physics

Undoubtedly, physicists have come to realize by now that theoretical concepts, particularly mathematical ones, are necessary in physics. However, physics does not need theories that are ambiguous and cannot be verified by an experiment within their reach. General relativity theory and the relativistic cosmology stood up positively to all experimental tests up to now. It seems to be premature to consider theories which try to refine general relativity without experimental motivation. Nevertheless, G.S. Asanov [13] comments that the Finslerian approach is nowhere found as an alternative to the Riemannian general or special theory of relativity. Indeed, it leads to their extension, developing their methods and bringing forth a new concept of space-time. According to Antonelli, Ingarden and Matsumoto [7], there are some examples indicating that Finsler geometrical models can be non-universal, depending upon the value e / m = electric charge / mass. They depend upon arbitrary gauge functions. For distinct values of e / m and for various gauges there exist different Finsler spaces. In the philosophy of physics, non-Riemannian models of differential geometry may be useful as approximate, "higher level" descriptions of composite systems and, in particular, for hierarchical systems of chemistry, macroscopic mechanics and biology. Human bodies are built of atoms, but for life, consciousness, etc. depend directly on concrete atoms, since they are constantly in flow in metabolic processes but it only depends on their structure as form which is preserved.

To quote some deeper physical interpretation of Finsler geometry, let us imagine the element of support of Elié Cartan as polarized photons or polarized electrons (for other material particles with non-zero spin). Then the extended Finsler-Cartan space is appropriate. However, when we deal with non-polarized photons or non-polarized electron beams and the direction of spin is suppressed by a uniform probability distribution, then the picture of a point in non-extended Finsler space is the right one. The directional statistics of spins have nothing to do with the directional properties of the electromagnetic field in which the particles move, so anisotropic metrics of Finsler geometry are necessarily useful.

Symplectic spaces are very important for mechanics (classical as well as quantum) together with its complex presentations. The closely connected contact spaces, except in mechanics and optics, also find their applications in phenomenological thermodynamics (Hermann [47], Mrugala [154], Mrugala et al. [155]). Affine connection spaces with their applications to general relativity with angular momentum (Einstein-Cartan spaces) and gauge theories of elementary particles are also represented (Binz, Sniatycki and Fischer [22], Dorodzinski [38]).

There are many applications of Finslerian approach to physics and engineering of materials, especially elasticity, ferromagnetism and other properties of oriented bodies (Ericksen and Truesdell [42], Truesdell and Toupin [225], Amari [5], Kondo and Amari [60] and Bejancu [17]).

§ 3. Mechanics of open systems

A classical mechanical system having all internal forces, i.e. forces of mutual interactions of particles of the system, is called an *isolated system* or a *closed system*. Such a system is conservative, i.e. its momentum, angular momentum and energy are independent of time. Also, systems that are non-isolated, i.e. open, but with all external forces independent of time are conservative. No external force can be absolutely constant, because of the Newton's laws of action and reaction. In this category one can include systems with a finite number of integrable constraints (holonomic) and independent of time. When external forces depend upon time, i.e. when constraints are rhenomic, energy is no longer conserved (Hamiltonian and Lagrangian depend explicitly on time) and the system can be called *dissipative*. In particular, when energy monotonically decreases in time, the system can be called *passive*; when energy monotonically increases in time, the system can be called *pumped* or *active*. The time evolution of a conservative system forms a group called a *dynamical group*. In the case of dissipative system, the time evolution forms a semi-group called a *dynamical semi-group*. In particular, a statistical quantum dynamical semi-group was first defined by A. Kossakowski [61].

The geometrical background of conservative dynamical systems (in the sense of the Hamiltonian variational principle) is symplectic geometry, and (in the sense of the Fermat-type principle) Riemannian or Finslerian geometry. For dissipative systems, both deterministic and statistical (also in thermodynamical phenomenological approach), such a background is formed by contact geometry (Mrugala [154], Mrugala et al. [155]). Finsler geometry can also be introduced for control dynamical systems. A geometrical method has been elaborated connecting mostly non-holonomic coordinate systems and connection (Neymark and Furfayev [156], Vershik and Greshkovich [229]). As regards rhenomic systems, A. Wundheiler [232] introduced the concept of rhenomic geometry, being a generalization of the Riemannian geometry (coordinate transformations depending upon time).

§ 4. Thermodynamic Finsler spaces

Using the point of view of statistical thermodynamics interpreted by means of information theory, Ingarden and Nakagomi [51] introduced *thermodynamical Finslerian geometry*. Their approach consists in maximizing entropy (information) of the probability distribution on the phase of a physical system by some number of given mean values has macroscopically measurable and are the only essential thermodynamical parameters together with the conjugated Lagrange multipliers. For instance, mean energy, mean volume, mean number of particles as mean values together with temperature and chemical potential as Lagrange multipliers. The maximization procedure gives the well-known exponential expressions for the equilibrium states as the *Gibbs states* and their generalizations. The deterministic mechanical states are the points $w = (w^1, \ldots, w^{2d})$ of the phase space of a holonomic mechanical system with $d < \infty$ degree of freedom. The phase space W is a smooth measurable $2d$-dimensional symplectic space, assuming the symplectic coordinates and a Liouville measure m invariant with respect to the canonical transformations. We assume that a thermodynamical state μ_x (a generalized Gibbs state) depending upon thermodynamical parameters $x = (x^1,\ldots, x^n)$ is a probability measure on W absolutely continuous with respect to m, $\mu \leq m$. Then there exists a unique Radon-Nikodym derivative called *probability distribution* or *probability density*. Now, we define relative entropy of state μ_x (or briefly x') with respect to state x, also called Kullback-Leibler [63] information, directed divergence, information gain or variation of information (Guiasu [45]).

A Finsler space is locally Minkowskian and therefore flat. Physically, this is caused by vanishing interactions between particles in the ideal classical gas. In quantum gases of *fermions* and *bosons*, the interaction cannot be completely eliminated and we get curved spaces: for fermions with a negative curvature (Ingarden et al. [50]); and for bosons with a positive curvature (Janyszek and Mrugala [54]). Conclusively, in crystal optics, the Finsler distance corresponds to the phase of an optical wave. In physiological optics, the Finsler distance corresponds to the usual geometrical length. In mechanics, in the Fermat-Jacobi problem, the Finsler distance has dimension of energy while in thermodynamics that of entropy. Thus, in all these cases, the Finsler distance has a direct physical interpretation based on experience.

[38)] Special unitary

BIBLIOGRAPHY

[1] Abate, M. and Patrizio, G.: *Finsler metrics - a global approach, with applications to geometric function theory.* Springer-Verlag, Berlin, Heidelberg, 1994.

[2] Agrawal, Pushpa: On the concircular geometry in Finsler spaces. *Tensor (N.S.)* 23 (1972), 333-336.

[3] Akbar-Zadeh, H. and Wegrzynowska, A.: Sur la géométrie du fibré tangent à une variété finslérienne. *C. R. Acad. Sci. Paris* A 282 (1976), 325-328.

[4] Al-Borney, M.S.: On the angular metric in areal spaces. *Tensor (N.S.)* 25 (1972), 372-374.

[5] Amari, S.: *Differential Geometrical Methods in Statistics*, Lecture notes in Statistics, vol. 28, Springer, Berlin, 1985.

[6] Antonelli, P. and Bradbury, R.: *Volterra-hamilton Models in Ecology and Evolution of Colonial Organisms*, World Scientific Press, 1994.

[7] __; Ingarden, R.S. and Matsumoto, M.: *The theory of sprays and Finsler spaces with applications in physics and biology.* Kluwer Academic Publishers, 1993.

[8] __ and Lackey, B.: *Finslerian Laplacians and Applications.* Kluwer Press, 1998.

[9] __ and Miron, R.: *Lagrange and Finsler Geometry Applications to Physics and Biology.* Kluwer Academic Press, 1996.

[10] __ and Zastawniak, T.: *Lagrange Geometry, Finsler Spaces and Noise Applied in Biology and Physics.* Pergamon Press, Mathematical and Computer Modelling, vol. 20, no. 4/5, 1994.

[11] __ and Zastawniak, T.: *Fundamentals of Finslerian Diffusion with Applications.* Kluwer Academic Press, 1998.

[12] Asanov, G.S.: New examples of S_3-like Finsler spaces. *Rep. on Math. Phys.* 16 (1979), 329-333.

[13] __: *Finsler Geometry, Relativity and Gauge Theories*. D. Reidel Publishing Co, Dordrecht (Holland), 1985.

[14] Awasthi, G.D.: A study of certain special Finsler spaces. *Univ. Nac. Tucumán Rev. Ser. A* 24 (1974/75), 163-166.

[15] Bao, D.; Chern, S. S. and Shen, Z.: *Finsler geometry*. A.M.S. Contemporary Mathematics, vol. 196, 1996.

[16] __: *An Introduction to Riemann-Finsler Geometry*. Spring-Verlag, 2000.

[17] Bejancu, A.: *Finsler geometry and applications*. Ellis Horwood Ltd., Chichester, 1990.

[18] Berwald, L.: Über Parallelübertragung in Räumen mit allgemeiner massbestimmung. *Jber. Deutsch. Math.-Verein* 34 (1926), 213-220.

[19] __: Untersuchung der Krümmung allgemeiner metrisher Räume auf Grund des in ihnen herrschenden Parallelismus. *Math. Z.* 25 (1926), 40-73.

[20] __: Parallelübertragung in allgemeiner Räumen. *Atti Congr. Bologna* 4 (1931), 263-270.

[21] __: Über Finslersche und Cartansche Geometrie, III. Two-dimensional Finsler spaces with rectilinear extremals. *Ann. Math.* (2) 42 (1941), 84-112.

[22] Binz, E.; Sniatycki, J. and Fischer, H.: *Geometry of Classical Fields*. North-Holland Pubg. Co., Amsterdam, 1988.

[23] Blumenthal, L.M.: *Theory and Applications of Distance Geometry*. Oxford Press, London, 1953.

[24] Brickell, F.: On areal spaces. *Tensor* (*N.S.*) 13 (1961), 19-30.

[25] __ and Al-Borney, M.S.: A note on areal measures. *J. London Math. Soc.* (2) 4 (1972), 466-468.

[26] Busemann, H.: Über die Geometrien, in dennen die Kreise mit unendlichem Radius die Kürzesten Linien sind. *Math. Ann.* 106 (1932), 140-160.

[27] __: Über die Räume mit konvexen Kugeln und Parallelenaxiom. *Nachr. Ges. Wiss. Göttingen* (1933), 116-140.

[28] __: *Metric methods in Finsler spaces and the Foundations of Geometry.* Ann. Math. Studies 8, Princeton, 1942.

[29] __: Foundations of Minkowskian Geometry. *Comm.Math. Helv.* 24 (1950), 156-187.

[30] __: *The Geometry of Geodesics.* New York, 1955.

[31] Cartan, Elié: Sur une classe remarquable d'espaces de Riemann. *Bull. Soc. Math. France* 54 (1926), 214-264.

[32] __: Sur les espaces de Finsler. *C.R. Acad. Sci. Paris* 196 (1933), 582-586.

[33] __: *Les Espaces de Finsler.* Actualites Scientifiques et Industrielles no. 79, Paris, Hermann, 1934.

[34] D' Alembert: *Mélange's de Littérature*, Amsterdam, 1759.

[35] Davies, E.T.: Lie derivation in generalized metric spaces. *Ann. Mat. Pura Appl.* 18 (1939), 261-274.

[36] __: On the notion of Euclidean connection in areal spaces. *Tensor (N.S.)* 24 (1972), 53-55.

[37] Delens, P.C.: *La metrique angulaire des espaces de Finsler.* Paris, Hermann, 1934.

[38] Dorodzinski, A.: *Geometry of the Standard Model of the Elementary Particles.* Berlin, 1991.

[39] Douglas, J.: The general geometry of paths. *Ann. Math.* (2) 29 (1928), 143-168.

[40] Drechsler, W. and Mayer, M. E.: *Fibre Bundle Techniques in Gauge Theories*. Springer, Berlin, 1977.

[41] Eisenhart, L.P.: *Riemannian Geometry*. Princeton Univ. Press, 1st ed. 1926, reprinted 1997.

[42] Ericksen, J.L. and Truesdell, C.: Exact theory of stress and strain in rods and shells. *Arch. Ration. Mech. Anal.* 1 (1958), 295-323.

[43] Finsler, P.: *Über Kurven und Flächen in allgemeinen Räumen*, Dissertation, Göttingen, 1918, published by Verlag Birkhäuser, Basel, 1951.

[44] Gauss, G.F. (1824): Letter to Taurinus, 8th November.

[45] Guiasu, S.: *Information Theory with Applications*. Mc-Graw Hill, New York, 1977.

[46] Gupta, B.: On projective symmetric spaces. *J. Austral. Math. Soc.* 4 (1964), 113-121.

[47] Hermann, R.: *Geometry, Physics and Systems*. Marcel Dekker, New York, 1973.

[48] Hiramatu, H.: Groups of homothetic transformations in a Finsler space. *Tensor (N.S.)* 3 (1954), 131-143.

[49] __: On some properties of groups of homothetic transformations in Riemannian and Finslerian spaces. *Tensor (N.S.)* 4 (1954), 28-39.

[50] Ingarden, R.S., Janyszek, H., Kossakowski, A. and Kawaguchi, T.: Information geometry of quantum statistical systems. *Tensor (N. S.)* 37, 1985.

[51] __ and Nakagomi, T.: The second order extension of the Gibbs state. *Open Syst. Inform. Dyn.* 1 (1992), pp. 243-258.

[52] Izumi, H.: Conformal transformations of Finsler spaces - I: Concircular transformations of a curve with Finsler metric. *Tensor (N.S.)* 31 (1977), 33-41.

[53] __: Conformal transformations of Finsler spaces - II: An *h*-conformally flat Finsler space. *Tensor (N.S.)* 34 (1980), 337-359.

[54] Janyszek, H., and Mrugala, R.: *Riemannian and Finslerian geometry and fluctuations of thermodynamical systems*. Advances in Thermodynamics vol. 3, Taylor and Francis. New York, 1990, pp. 159-174.

[55] Katzin, G.H.; Levine, Jack and Davis, W.R.: Curvature collineations: A fundamental symmetry property of the space-times of general relativity defined by the vainishing of Riemann curvature tensor. *J. Math. Phys.* 10 (1969), 617-629.

[56] Kawaguchi, A.: Geometry of an *n*-dimensional space with the arc-length $s = \int (A_i x''^{\,i} + B)^{1/p}$. *Trans. Amer. Math. Soc.* 44 (1938), 153-167.

[57] Kawaguchi, Syun-ichi: Some properties of the curvature tensors in special Kawaguchi space of even dimensions. *Tensor (N.S.)* 25 (1972), 451-454.

[58] Klein, F. (1873): Uber die sogenannte Nicht-Euklidische Geometrie, *Math. Annalen* 6.

[59] Knebelman, M.S.: Conformal geometry of generalized metric spaces. *Proc. Nat. Acad. Sci.* (*USA*) 15 (1929), 376-379.

[60] Kondo, K. and Amari, S.: A constructive approach to the non-Riemannian features of dislocation and spin distributions in terms of Finsler Geometry and a possible extension to the space-time formation. *RAAG Memoirs* 4D (1968), 225-238.

[61] Kossakowski, A.: On quantum statistical mechanics of non-Hamiltonian systems. *Rep. Math. Phys.* 3 (1972), pp. 247-274.

[62] Kropina, V.K.: On projective Finsler spaces with a certain special form. *Naučn. Doklady vyss. Skoly, fiz.-mat. Nauki* 2 (1959 / 60), 38-42 (Russian).

[63] Kullback, S. and Leibler, R.A.: On information and sufficiency. *Ann. Math. Statist.* 22 (1951), pp. 79-86.

[64] Kumar, A.: On special projective tensor fields. *Atti Accad. Naz. Lincei Rend. Cl. Sci. Fis. Mat. Natur.* (8) 58 (1975), 184-189.

[65] __: On the existence of affine motion in a recurrent Finsler space. *Indian J. Pure Appl. Math.* 8 (1977), 791-800.

[66] Lambert, J.H. (1786): Theories de Parallelinien I. *Leipziger Magazin für Rheine und angewandte Mathematik* 1 (2) (1786), 137-164; II. *Ibid* 1 (3) (1786), 325-358.

[67] Lichnerowicz, André: *Théories relativistes de la gravitation et de l'electromagnetisme*. Paris, 1955.

[68] Lobachevsky, N.I. (1891): Geometrische Untersuchungen, English translation by G.B. Halsted.

[69] Ludlam, W. (1785): *The Rudiments of Mathematics*. Cambridge, 1785.

[70] Matsumoto, Makoto: Affine transformations of Finsler spaces. *J. Math. Kyoto Uni.* 3 (1963), 1-35.

[71] __: A Finsler connection with many torsions. *Tensor* (*N.S.*) 17 (1966), 217-226.

[72] __: On Riemannian spaces with recurrent projective curvature. *Tensor* (*N.S.*) 19 (1968), 11-18.

[73] __: *The Theory of Finsler Connections*. Publ. Study Group of Geometry 5, Okayama Univ., 1970.

[74] __: On Finsler spaces with curvature tensors of some special forms. *Tensor* (*N.S.*) 22 (1971), 201-204.

[75] __: On *C*-reducible Finsler spaces. *Tensor* (*N.S.*) 24 (1972), 29-37.

[76] __: *V*-transformations of Finsler spaces - I: Definition, infinitesimal transformations and isometries. *J. Math. Kyoto Univ.* 12 (1972), 479-512.

[77] __: On Finsler spaces with Rander's metric and special forms of important tensors. *J. Math. Kyoto Univ.* 14 (1974), 477-498.

[78] __: Finsler spaces admitting a concurrent vector field. *Tensor* (*N.S.*) 28 (1974), 239-249.

[79] __: On three-dimensional Finsler spaces satisfying the *T*- and B^p - conditions. *Tensor* (*N.S.*) 29 (1975), 13-20.

[80] __: Finsler spaces with *hv*-curvature tensor $P_{h\,i\,j\,k}$ of a special form. *Rep. on Math. Phys.* 14, (1978), 1-13.

[81] __: Fundamental functions of S_3-like Finsler spaces. *Tensor* (*N.S.*) 34 (1980), 141-146.

[82] __: Theory of extended point transformations of Finsler spaces - I. Fundamental theorems of affine motions. *Tensor* (*N.S.*) 34 (1980), 303-315.

[83] __: *Foundations of Finsler Geometry and Special Finsler Spaces*. Kaeseisha Press, Saikawa, Otsu 520, 1986.

[84] __: Theory of extended point transformations of Finsler spaces - II. Fundamental theorems of projective motions. *Tensor* (*N.S.*) 47 (1988), 203-214.

[85] __ and Eguchi, K.: Finsler spaces admitting a concurrent vector field. *Tensor* (*N.S.*) 28 (1974), 239-249.

[86] Meetz, K.: Realization of chiral symmetry in a curved isotopic space. *J. Math. Phys.* 10 (1969), pp. 589-593.

[87] Meher, F.M.: *Certain investigations in Finsler spaces*. Ph. D. Thesis, Berhampur-Ganjam (India), 1972.

[88] __: Projective motion in a symmetric Finsler space. *Tensor* (*N.S.*) 23 (1972), 275-278.

[89] __: Projectively flat manifolds with recurrent curvature. *Boll. Un. Mat. Ital. B* (5), 15-B (1978), 828-834.

[90] Miron, R.: *The geometry of higher-order Finsler spaces*. Hadronic Press, Inn., USA, 1998 (ISBN 1-57485-033-4).

[91] __ and Anastasiei, M.: *Vector bundles and Lagrange spaces with applications to relativity, Geometry*. Balkan Press, Bucharest, Romania, 1997.

[92] Mishra, R.S.: *A course in Tensors with application to Riemannian geometry*. Pothishala Pvt. Ltd., Allahabad, 1965.

[93] __: *Structures on a Differentiable Manifold and Their Applications*. Chandrama Prakashan, Allahabad, 1984, ASIN: B0007BFQ GA.

[94] __: *Almost Contact Metric Manifolds*. Tensor Society of India, Lucknow, Monographs vol. 1, 1994.

[95] __ and Misra, R.B.: The Killing vector and the generalized Killing equation in Finsler space. *Rend. Circ. Mat. Palermo* (2) 15 (1966), 216-222; MR 37 # 3488; Zbl. 168 # 432.

[96] __, __ and Nawal-Kishore: On a symmetric Finsler manifold admitting an affine motion. *Bull. Soc. Math. Belgique Sér A* 30 (1978), 39-44; MR 84 a # 53028; Zbl. 484 # 53020.

[97] __ and Pande, H.D.: Recurrent Finsler spaces. *J. Indian Math. Soc.* 32 (1968), 17-22.

[98] Misra, R.B.: The projective transformation in a Finsler space. *Ann. Soc. Sci. Bruxelles Sér. I* 80 (1966), 227-239; MR 34 # 6704; Zbl. 154 # 218.

[99] __: Projective tensors in a conformal Finsler space. *Acad. Roy. Belgique Bull. Cl. Sci.* (5) 52 (1966), 1275-1279; MR 35 # 6112; Zbl. 154 # 218.

[100] __: The commutation formulae in a Finsler space, I. *Ann. Mat. Pura Appl.* (4) 75 (1967), 363-370; MR 35 # 4866; Zbl. 145 # 422.

[101] __: The commutation formulae in a Finsler space, II. *Ann. Mat. Pura Appl.* (4) 75 (1967), 371-384; MR 35 # 4866; Zbl. 145 # 422.

[102] __: The Bianchi identities satisfied by curvature tensors in a conformal Finsler space. *Tensor* (*N.S.*) 18 (1967), 187-190; MR 36 # 2098; Zbl. 147 # 217.

[103] __: *Some Problems in Finsler spaces.* D. Phil. Thesis, Allahabad (India), 1967.

[104] __: On the deformed Finsler space. *Tensor* (*N.S.*) 19 (1968), 241-250; MR 38 # 5145; Zbl. 162 # 257.

[105] __: The generalized Killing equation in Finsler space. *Rend. Circ. Mat. Palermo* (2) 18 (1969), 99-102; MR 43 # 6854; Zbl. 231 # 53048.

[106] __: Projective invariants in a conformal Finsler space. *Tensor* (*N.S.*) 21 (1970), 186-188; MR 41 # 9167; Zbl. 191 # 203.

[107] __: On the generalized Lie differentiation arising from Su's infinitesimal transformation. *İstanbul Üniv. Fen Fak. Mecm. Ser. A* 35 (1970), 5-15; MR 49 # 6080; Zbl. 254 # 53011.

[108] __: A symmetric Finsler space. *Tensor* (*N.S.*) 24 (1972), 346-350; MR 48 # 12436; Zbl. 232 # 53034.

[109] __: A projectively symmetric Finsler space. *Math. Zeit.* 126 (1972), 143-153; MR 45 # 7662; Zbl. 232 # 53033.

[110] __: On a recurrent Finsler space. *Rev. Roumaine Math. Pures Appl.* 18 (1973), 701-712; MR 48 # 1086; Zbl. 259 # 53028.

[111] __: A bi-recurrent Finsler manifold with affine motion. *Indian J. Pure Appl. Math.* 6 (1975), 1441-1448; MR 58 # 30883; Zbl. 369 # 53031.

[112] __: A turning point in the theory of recurrent Finsler manifolds. *J. South Gujarat Univ.* 6 (1977), 72-96; MR 80 a # 53032; Zbl. 404 # 53020.

[113] __: A turning point in the theory of recurrent Finsler manifolds, II. Certain types of projective motions. *Boll. Un. Mat. Ital.* (5) 16-B (1979), 32-53; MR 81 i # 53025; Zbl. 413 # 53013.

[114] __: Groups of transformations in Finslerian spaces. *International Centre for Theoretical Phys., Trieste* IC/81/241 (1981), 1-15 (Preprint).

[115] __: *Basic Concepts of Finslerian Geometry*. International Centre for Theoretical Physics, Trieste IC/86/278 (1986), 1-37 (Preprint).

[116] __: Groups of transformations in Finslerian spaces (revised). *I.C.T.P., Trieste* IC / 93 / 11 (1993), 1-19 (Preprint).

[117] __: *Differential Geometry of Curves & Surfaces*. Hardwari Publications, Allahabad (India), 2002.

[118] __: *Tensors*. Ibid, 2002; MR 2003 d # 53022

[119] __: *On projectively flat Finslerian spaces*. Lambert Academic Publishing House, Saarbrucken (Germany), pp. 2-22, 2011, ISBN 978-3-8443-0037-6.

[120] __: *On Finsler spaces with concircular transformations*. Ibid, 2011, ISBN 978-3-8443-0037-6.

[121] __: *On Finsler spaces with concircular transformations II*. Ibid, ISBN 978-3-8443-0037-6, 2011.

[122] __: *Glossary of Mathematical Terms and Concepts - Part* 1, Central West Publishing, Orange, NSW 2800 (Australia), 2019, ISBN (print): 978-1-925823-68-4, ISBN (ebook): 978-1-925823-69-1.

[123] __: *Glossary of Mathematical Terms and Concepts - Part* 3, Ibid, 2019, ISBN (print): 978-1-925823-73-8.

[124] __: *Glossary of Mathematical Terms and Concepts, Part* 4, Ibid, 2019, ISBN (print): 978-1-925823-74-5.

[125] __ and Ahmad, S.M.W.: Gauge unification of fundamental forces: the story of success. *I.C.T.P., Trieste* IC/93/12 (1993), 1-13 (Pre-print); *J. of International Acad. of Physical Sci., Allahabad* 1 (1997), 1-11.

[126] __ and Fava, Franco: Eulerian curvature tensors and the conformal mapping. *Rend. Sem. Mat. Univ. e Politec. Torino* 35 (1976/77), 311-326; MR 58 # 12802; Zbl. 375 # 53013.

[127] __ and Meher, F.M.: Projective motion in an *RNP*-Finsler space. *Tensor (N.S.)* 22 (1971), 117-120; MR 43 # 5471; Zbl. 206 # 509.

[128] __, __: A *SHR*-F_n admitting an affine motion. *Acta Math. Acad. Sci. Hungar.* 22 (1971), 423-429; MR 45 # 2637; Zbl. 238 # 53031.

[129] __, __: Some commutation formulae arising from Lie differentiation in a Finsler space. *Atti. Accad. Naz. Lincei Rend. Cl. Sci. Fis. Mat. Natur.* (8) 50 (1971), 18-23; MR 46 # 816; Zbl. 215 # 512.

[130] __, __: On the existence of affine motion in a *HR*- F_n . *Indian J. Pure Appl. Math.* 3 (1972), 219-225; MR 45 # 5932; Zbl. 237 # 53022.

[131] __, __: Lie differentiation and projective motion in the projective Finsler space. *Tensor (N.S.)* 23 (1972), 57-65; MR 46 # 6219; Zbl. 228 # 53016.

[132] __, __: A Finsler space with special concircular projective motion. *Tensor (N.S.)* 24 (1972), 288-292; MR 48 # 9595; Zbl. 232 # 53032.

[133] __, __: A recurrent Finsler space of second order. *Rev. Roumaine Math. Pures Appl.* 18 (1973), 563-569; MR 47 # 5764; Zbl. 253 # 53025.

[134] __, __: *CA*-motion in a *PS*-F_n. *Indian J. Pure Appl. Math.* 6 (1975), 522-526; MR 57 # 7438; Zbl. 373 # 53012.

[135] __, __ and Nawal-Kishore: On a recurrent Finsler manifold with a concircular vector field. *Acta Math. Acad. Sci. Hungar.* 32 (1978), 287-292; MR 80 a # 53033; Zbl. 413 # 53015.

[136] Misra, R.B. and Mishra, C.K.: Infinitesimal deformation of curves in a Finsler space. *J. of International Acad. of Physical Sci.* (2001).

[137] __, __: Torse-forming infinitesimal transformations in a Finsler space. *Tensor (N.S.)* 65 (2004), 1-7.

[138] __ and Mishra, R.S.: Lie derivatives of various geometric entities in Finsler Space. *İstanbul Üniv. Fen Fak. Mecm. Ser. A* 30 (1965), 77-82; MR 39 # 7548; Zbl. 172 # 233.

[139] __, __ and Nawal-Kishore: On bisymmetric Finsler manifolds. *Boll. Un. Mat. Ital.* (5) 14-A (1977), 157-164; MR 57 # 10633; Zbl. 344 # 53017.

[140] Misra, R.B. and Nawal-Kishore: On curvature collineations in Finsler manifolds. *Atti Accad. Sci. Lett. Arti Palermo Parte I* (4) 36 (1976/77), 521-534; MR 81 b # 53025; Zbl. 468 # 53018.

[141] __ , __and Pandey, P.N.: Projective motion in an *SNP-*F_n . *Boll. Un. Mat. Ital.* (5) 14-A (1977), 513-519; MR 57 # 17543; Zbl. 388 # 53023.

[142] __ and Pande, K.S.: On the Finsler space admitting a holonomy group. *Ann. Mat. Pura Appl.* (4) 85 (1970), 327-346; MR 41 # 6138; Zbl. 194 # 535.

[143] __, __: On Misra's covariant differentiation in a Finsler space. *Atti Accad. Naz. Lincei Rend. Cl. Sci. Fis. Mat. Natur.* (8) 48 (1970), 199-204; MR 42 # 8434; Zbl. 195 # 239.

[144] __ and Pandey, P.N.: Projective recurrent Finsler manifolds, I. *Publ. Math. Debrecen* 28 (1981), 191-198; MR 83 b # 53027; Zbl. 484 # 53023.

[145] Moór, A.: Entwicklung einer Geometrie der allgemeinen metrischen Linienelementräume. *Acta Sci. Math.* (*Szeged*) 17 (1956), 85-120.

[146] __: Eine Verallgemeinerung der metrischer Übertragung in allgemeinen metrischen Räumen. *Publ. Math. Debrecen* 10 (1963), 145-150.

[147] __: Untersuchungen über Finslerräume von rekurrenter Krümmung. *Tensor* (*N.S.*) 13 (1963), 1-18.

[148] __: Über Finslerräume von Zweifach rekurrenter Krümmung. *Acta Math. Acad. Sci. Hungar.* 22 (1971), 453-465.

[149] __: Ünterräume von rekurrenter Krümmung in Finslerräumen. *Tensor* (*N.S.*) 24 (1972), 261-265.

[150] __: Finslerräume von rekurrenter Torsion. *Publ. Math. Debrecen* 21 (1974), 255-265.

[151] __: Über eine Übertragungstheorie der metrischen Linienelementräume mit rekurrenter Grundtensor. *Tensor* (*N.S.*) 29 (1975), 47-63.

[152] __: Übertragungstheorien in Finslerschen und verwandten Räumen. *Berichte Math.-Stat. Sekt. Forschungzentrum, Graz* 42 (1975), 1-14.

[153] __: Über die Veranderung der Krümmung bei einer Torsion der Übertragung. *Acta Math. Acad. Sci. Hungar.* 26 (1975), 97-111.

[154] Mrugala, R.: *Contact and Metric Geometry in Thermodynamics* (in Polish). Nicholas Copernicus Univ. Press, Torun, 1990.

[155] __, Nulton, J.D., Schon, J. Ch. and Salamon, P.: Contact structures in thermodynamic theory. *Rep. Math. Phys.* 29 (1985), pp. 109-121.

[156] Neymark, Yu. I. and Furfayev, N.A.: *Dynamics of non-holonomic systems* (in Russian). Nauka, Moscow, 1967.

[157] Noether, E.: Invarianten beliebiger Differentialausdrücke. *Gött. Nachr.* 25 (1918), 37-44.

[158] Nomizu, K.: On hypersurfaces satisfying a certain condition on the curvature tensor. *Tôhuku Math. J.* 20 (1968), 46-59.

[159] Numata, S.: On Landsberg spaces of scalar curvature. *J. Korean Math. Soc.* 12 (1975), 97-100.

[160] __: On the torsion tensors $R_{h\,j\,k}$ and $P_{h\,j\,k}$ of Finsler spaces with a metric $(ds)^2 = g_{i\,j}\,(x)\,dx^i\,dx^j + b_i\,(x)\,dx^i$. *Tensor* (*N.S.*) 32 (1978), 27-31.

[161] Okumura, M.: On some types of connected spaces with concircular vector fields. *Tensor* (*N.S.*) 12 (1962), 33-46.

[162] Pande, H.D.: The projective transformation in Finsler space. *Atti Acad. Naz. Lincei Cl. Sci. Fis. Mat. Natur.* (8) 43 (1967), 480-484.

[163] __: Various Ricci identities in Finsler space. *J. Austral. Math. Soc.* 9 (1969), 228-232.

[164] __: Lie derivation in a Minkowskian Finsler space. *Atti Acad. Naz. Lincei Cl. Sci. Fis. Mat. Natur.* (8) 50 (1971), 699-702.

[165] __: *A study of Some Problems in Finsler Space*. Pustaksthan, Buxipur, Gorakhpur (India), 1974.

[166] __ and Kumar, A.: Infinitesimal special projective transformation in Finsler space. *J. Pure Appl. Math. Sci.* 2 (1975), 11-15.

[167] __, Kumar, A. and Khan, T.A.: Curvature collineations in a Finsler space. *Acta Ci. Indica* 1 (1975), 357-360.

[168] __ and Pandey, J.P.: A *W*-recurrent Finsler space of the second order. *Bull. de l'Acad. Serbe des Sci. et des Arts, Cl. des Sci. Math. et Natur.* 55 (1976), 65-71.

[169] __and __: Some theorems on projective motion in a special symmetric Finsler space. *İstanbul Üniv. Fen Fak. Mecm. Ser. A* 46/47 (1981/82), 55-60.

[170] Pandey, P.N.: Contra projective motion in a Finsler manifold. *Math. Education* 11 (1977), 25-29.

[171] __: Conformal covariant derivative in a Finsler manifold. *Atti Accad. Sci. Lett. Arti Palermo Parte I* (4) 37 (1977/78), 341-350.

[172] __: *CA*-collineation in a bi-recurrent Finsler manifold. *Tamkang J. Math.* 9 (1978), 79-81.

[173] __: Groups of conformal transformations in conformally related Finsler manifold. *Atti Accad. Naz. Lincei Rend. Cl. Sci. Fis. Mat. Natur.* (9) 65 (1978), 269-274.

[174] __: A recurrent Finsler manifold admitting special recurrent transformations. *Progr. Math., Allahabad* (*India*) 13 (1979), 85-98.

[175] __: A recurrent Finsler manifold with a concircular vector field. *Acta Math. Acad. Sci. Hungar.* 35 (1980), 465-466.

[176] __: Affine motion in recurrent Finsler manifold. *Ann. Fac. Sci. Univ. Nat. Zaire* (*Kinshasa*) *Sect. Math.-Phys.* 6 (1980), 51-63.

[177] __: A note on recurrence vector, *Proc. Nat. Acad. Sci.* (*India*) 51 (1981), 6-8.

[178] __: On a Finsler space of zero projective curvature. *Acta Math. Acad. Sci. Hungar.* 39 (1982), 387-388.

[179] __: A symmetric Finsler manifold with a concircular vector field, *Proc. Nat. Acad. Sci.* (*India*) 54-3 (1984), 271-273.

[180] __: Projective motions in a symmetric and projectively symmetric Finsler manifold. *Ibid*, 54-3 (1984), 274-278.

[181] __: Certain types of affine motion in Finsler manifold, III. *Colloq. Math.* 56 (1988), 333-340.

[182] __ and Dwivedi, V.J.: Normal projective curvature tensor in a conformal Finsler manifold. *Proc. Nat. Acad. Sci. India Sec. A* 47 (1977), 115-118.

[183] Prasad, B.N.: Curvature collineations in a Finsler space. *Atti Acad. Naz. Lincei Cl. Sci. Fis. Mat. Natur.* (8) 49 (1970), 194-197.

[184] Rani, N.: Theory of Lie derivatives in a Minkowskian space. *Tensor (N.S.)* 20 (1969), 100-102.

[185] Riemann, B.: Über die Hypothesen, welche der Geometrie zu Grunde liegen, Göttingen, 1854.

[186] Rund, H.: *Finsler spaces considered as locally Minkowskian spaces*. Ph.D. Thesis, Capetown, 1950.

[187] __: Über die parallelverschiebung in Finslerschen Räumen. *Math. Z.* 54 (1951), 115-128.

[188] __: On the analytical properties of curvature tensors in Finsler spaces. *Math. Ann*, 127 (1954), 82-104.

[189] __: *The Differential Geometry of Finsler Spaces*. Springer, Berlin, 1959.

[190] Ruse, H.S.: On simply harmonic "kappa-spaces" of four dimensions. *Proc. London Math. Soc.* 50 (1948), 317-329.

[191] Saccheri, G. (1733): Euclid's ab omni naevo vindicatus, Milan, 1733.

[192] Sanini, A.: Derivazioni su distribuzioni e connessioni di Finsler. *Rend. Sem. Mat. Univ. Politecn. Torino* 31 (1972/73), 1-28.

[193] __: Su un tipo di struttura quasi hermitiana del fibrato tangente ad uno spazio di Finsler. *Rend. Sem. Mat. Univ. Politecn. Torino* 32 (1973/74), 303-316.

[194] Savile, Sir Henry: *Praelectiones*. Oxford, 1621.

[195] Schouten, J.A.: *Ricci Calculus*. 2nd ed. (English), Springer, Berlin, 1954.

[196] Sen, R.N.: Finsler spaces of recurrent curvature. *Tensor (N.S.)* 19 (1968), 291-299.

[197] Shamihoke, A.C.: Some properties of curvature tensors in a generalized Finsler space. *Tensor (N.S.)* 12 (1962), 97-109.

[198] Shen, Z.: *Differential Geometry of Spray and Finsler Spaces.* Kluwer Academic Publishers, 2001.

[199] __: *Lectures on Finsler Geometry.* World Scientific Publishers, 2001.

[200] Shibata, C.: On the curvature tensor $R_{h\,i\,j\,k}$ of Finsler spaces of scalar curvature. *Tensor (N.S.)* 32 (1978), 311-317.

[201] __: On Finsler spaces with Kropina metric. *Rep. on Math. Phys.*13 (1978), 117-128.

[202] __; Shimada, H.; Azuma, M. and Yasuda, Y.: On Finsler spaces with Randers' metric. *Tensor (N.S.)* 31 (1977), 219-226.

[203] Singh, O.P.: On the projective motion in a projective Finsler space of recurrent curvature. *Mat. Vesnik* 10 (25), (1973), 105-110.

[204] Singh, S.P.: On the curvature collineations in Finsler spaces. *Publ. de l'Inst. Math. Yugoslavia* 36 (50), (1984), 85-89.

[205] __: On the curvature collineations in Finsler spaces - II. *Tensor (N.S.)* 44 (1987), 113-117.

[206] Singh, U.P.: On the induced theory of Finsler hypersurfaces from the standpoint of non-linear connection. *Atti Acad. Naz. Lincei Cl. Sci. Fis. Mat. Natur.* (8) 53 (1972), 541-548.

[207] __ and Prasad, B.N.: Special curvature collineations in Finsler space. *Atti Acad. Naz. Lincei Cl. Sci. Fis. Mat. Natur.* (8) 50 (1971), 122-127.

[208] Sinha, B.B.: *Studies in Finsler Spaces.* Ph.D. Thesis. Gorakhpur (India), 1962.

[209] __: On projective mapping in a Finsler space. *Tensor (N.S.)* 22 (1971), 326-328.

[210] __ and Singh, S.P.: On recurrent spaces of second order in Finsler spaces. *Yokohama Math. J.* 18 (1970), 27-32.

[211] __, __: Recurrent spaces of second order. *Yokohama Math. J.* 19 (1971), 79-85.

[212] Sinha, R.S.: *Certain Investigations in Finsler Space*. Ph.D. Thesis. Gorakhpur (India), 1963.

[213] __: Affine motion in recurrent Finsler space. *Tensor* (*N.S.*) 20 (1969), 261-264.

[214] __: On projective motions in a Finsler space with recurrent curvature. *Tensor* (*N.S.*) 21 (1970), 124-126.

[215] __: Infinitesimal projective transformations in Finsler spaces. *Progr. Math., Allahabad* (*India*) 5 (1971), 30-34.

[216] Slebodzinski, W.: Sur les equations de Hamilton. *Bull. Acad. Roy. de Belgique* (5) 17 (1931), 864-870.

[217] Soós, Gy.: Über die geodätischen abbildungen von Riemannscher Räumen auf projecktiv-symmetrische Riemannsche Räume. *Acta Math. Acad. Sci. Hungar.* 9 (1958), 359-361.

[218] Su, Buchin: Geodesic deviation in generalized metric spaces. *Academia Sinica Sci. Record* 2 (1949), 220-226.

[219] Szabó, Z. I.: Structure theorems on Riemannian spaces satisfying $R(X, Y)R = 0$. *J. Diff. Geom.* 17 (1982), 531-582.

[220] Takano, K.: Affine motion in non-Riemannian K^*-spaces, I, II, III (with M. Okumura), IV, V, *Tensor* 11 (1961), 137-143, 161-173, 174-181, 245-253, 270-278.

[221] __: On projective motion in a space with recurrent curvature. *Tensor* 12 (1962), 28-32.

[222] __: On the existence of affine motion in a space with recurrent curvature II. *Tensor* 17 (1966), 212-216.

[223] Taurinus, F.A.: *Theorie der Parallelinien*, 1825.

[224] Tomonaga, Y.: Jacobi fields in a Finsler space. *TRU Math.* 5 (1969), 37-42.

[225] Truesdell, C., and Toupin, R.A.: *The Classical Field Theories.* Hab. Phys. Vol. 3 (1), Springer, Berlin, 1960, pp. 226-790.

[226] Vacaru, Sergiu: Finsler and Lagrange Geometries in Einstein and String Gravity. *Intl. Jr. of Geometric Methods in Modern Physics* 5, no. 4 (2008), 473-511.

[227] Vagner, V.V.: Über Berwaldsche Räume. *Rec. Math. Moscow (N.S.)* 3 (1938), 655-662.

[228] __: On generalized Berwald spaces. *C.R. (Doklady) Acad. Sci. URSS (N.S.)* 39 (1943), 3-5.

[229] Vershik, A. M. and Greshkovich, V. Ya.: *Non-holonomic dynamical systems, Geometry of distributions and variational problems. Dynamical Systems* VII. Enc. Math. Sci. 16, Springer, New York, 1992.

[230] Walker, A.G.: On Ruse's spaces of recurrent curvature. *Proc. London Math. Soc.* 52 (1951), 36-64.

[231] Willmore, T.J.: *An Introduction to Differential Geometry.* Oxford University Press, London, 1959; Clarendon Press, Oxford, 1972 (reprinted).

[232] Wundheiler, A.: Über die Variationsgleichungen für affine geodetische Linien und nicht holonome, nicht conservative, dynamische Systeme. *Prace. Mat.-Fiz. (Warszawa)* 38, 1931.

[233] Yano, K.: Concircular geometry, I, II, III, IV, V, *Proc. Imp. Acad. Tokyo* 16 (1940), 195-200, 354-360, 442-448, 505-511, 18 (1942), 446-451.

[234] __: *The Theory of Lie derivatives and Its Applications.* North-Holland Publishing Co., Amsterdam, 1957.

[235] __: On the tangent bundles of Finsler and Riemannian manifolds. *Rend. Circ. Mat. Palermo* (2) 12 (1963), 211-228.

[236] __: *Differential Geometry of Complex and Almost Complex Spaces.* Pergamon Press, 1965.

[237] __ and Muto, Y.: Homogeneous contact manifolds and almost Finsler manifolds. *Kodai Math. Sem. Rep.* 21 (1969), 16-45.

[238] __ and Okubo, T.: On tangent bundles with Sasakian metrics of Finslerian and Riemannian manifolds. *Ann. Mat. Pura Appl.* (4) 87 (1970), 137-162.

[239] Yasuda, Hiroshi: On transformations of Finsler spaces. *Tensor* (*N. S.*) 34 (1980), 316-326.

[240] Yasuda, H. and Shimada, H.: On Randers' spaces of scalar curvature. *Rep. on Math. Phys.* 11 (1977), 347-360.

SYMBOLS AND ABBREVIATIONS

§ 1. General notes about symbols

Latin small fonts (without any suffixes) are generally used for scalar quantities unless stated otherwise. Points, curves and surfaces are generally denoted by capital non-italic Latin letters. On the other hand, both small and big Latin letters with one lower (resp. upper) suffix denote the components of covariant (resp. contravariant) vectors. As exceptions, u^λ, x^i, z^i along with their barred counterparts $\bar{x}^i$, $\bar{z}^i$, are used to denote the respective coordinates. Further exceptions are E_2, E_n, F_n, R_n, V_2, V_n, etc. used for spaces of dimensions as per their lower suffixes but C^2 denotes the order of derivability of the function.

Likewise, these letters with more than one lower (resp. upper) suffix generally denote the components of covariant (resp. contravariant) tensors. Further, those with both lower and upper suffixes are used for mixed tensors. Allowing exceptions, the connection parameters (which are not tensors) are also denoted by the similar symbols. Occasionally, Greek counterparts with above specifications are also used similarly for scalars / vectors / tensors / connection parameters. Especially, generalized Christoffel symbols (in Finsler geometry) are denoted by suffixed Greek small letters $\gamma_{i\,k\,j}$ and γ^{i}_{jk}. Some additional Greek letters and mathematical symbols such as δ^k_i, ∂_i, ∇_i, Π^{i}_{jk}, $\widetilde{\Pi}^{i}_{jk}$, $\mathcal{P}_k$, $\mathfrak{B}_j$, £, §, $\Re$, $\otimes$, etc. are also used and listed below. The bold face letters with upper circumflex accent $\hat{\mathbf{i}}$, $\hat{\mathbf{j}}$ are used for unit vectors along the rectangular Cartesian coordinate axes. Greek small letters in bold face are also used for vectors especially in Chapter 14. The operations dot (.) and cross (×) are used to denote the inner and outer products of vectors respectively. Usual mathematical notations of derivation, integration, summation, etc. and (algebraic) operations are also used without any further explanation.

Special care should be observed to distinguish different transformed geometric entities as they are often denoted by placing a single bar above them for coordinate change, projective change and conformal change.

1.1. Latin symbols

Symbol	Used for	Reference
$(a^1, a^2, \ldots, a^n)$	an n-tuple forming a vector **a**	p. 210
AHR - F_n	HR-F_n admitting an affine motion	p. 119
AHR^2-F_n	bi-recurrent Finsler space with affine motion	p. 160
A_n*	non-Riemannian space of recurrent curvature	p. 150
A_{lm}	covariant skew-symmetric tensor	Eq. (8.1.2)
$A_{ijk}, \overline{A}_{ijk}$	symmetric covariant tensors in F_n and conformally deformed space $\overline{F}_n$	pp. 27, 61
$A^h_{ik}, \overline{A}^h_{ik}$	mixed tensors in F_n and conformally deformed $\overline{F}_n$	pp. 27, 61
$ASHR$-F_n	SHR-F_n admitting an affine motion	p. 140
$(b^1, b^2, \ldots, b^n)$	an n-tuple forming a vector **b**	p. 210
B^{im}	a symmetric contravariant tensor	p. 62
B_{kh}	second order covariant tensor (not associate to above tensor)	p. 168
C^2	derivability up to second order	p. 18
C_{ijk}, C^h_{ik}	symmetric covariant tensor and its associate tensor in F_n	p. 20
$\overline{C}_{ijk}, \overline{C}^h_{ik}$	" " in conformally deformed space $\overline{F}_n$	p. 61
CHR-F_n	recurrent Finsler space admitting a c.i.t.	p. 199
$d\mathbf{r}$	infinitesimal vector through 2 neighbouring points on a surface	p. 14
ds	infinitesimal arc-length of a curve	p. 8
dS	elementary region of a surface	p. 6
$du^\lambda, \delta u^\lambda$	small increments in the parameters u^λ	p. 14
$D_k X^i$	covariant derivative of vector field X^i	Eq. (4.5.1)
$\underset{L}{D}$	Rund's notation for Lie differential	p.78
$\{\mathbf{e}_i\}, \{\overline{\mathbf{e}}_i\}$	bases of a vector space V or V^1	p. 211

$\{\mathbf{e}^i\}$, $\{\bar{\mathbf{e}}^i\}$	bases of the dual vector space V_1	p. 212
$\mathbf{e}_{kh}$	bilinear functions over the space $V_1 \times V_1$	p. 214
$\mathbf{e}^{kh}$	" " over the product space $V^1 \times V^1$	p. 215
$\{\mathbf{e}^{j_1 j_2 \ldots j_s}_{i_1 i_2 \ldots i_r}\}$	base vectors of the vector space V^r_s	p. 218
E_2, E_3, E_n	Euclidean spaces of 2, 3, n-dimensions	p. 13
f_h	a covariant vector-field	Eq. (12.5.4)
f^{jk}	skew-symmetric tensor	p. 125
F	(i) metric function, and also used for (ii) non-null set of scalars forming an algebraic field	p. 8, p. 209
$\bar{F}$	metric function of conformally deformed space $\bar{F}_n$	p. 59
F_n	n-dimensional Finsler space	p. 20
$\bar{F}_n$	deformed space of F_n under (i) coordinate change $x^i \to \bar{x}^i$, also in (ii) projective change $G^i \to \bar{G}^i$, and (iii) under conformal change $g_{ij} \to \bar{g}_{ij}$	pp. 51, 59-60
g^2	discriminant of metric	Eq. (3.2.11)
$g_{\lambda\mu}$	first order fundamental magnitudes or components of metric tensor of a surface	Eq. (3.2.6)
$((g_{\lambda\mu}))$	matrix of coefficients in metric	Eq. (3.2.17)
g^{ij}, $g^{\lambda\mu}$	components of associate metric tensor	p. 22
$\bar{g}_{ij}$, $\bar{g}^{ij}$	metric tensor and its associate tensor in (i) conformally deformed space $\bar{F}_n$, also in (ii) infinitesimally deformed space $\bar{F}_n$	p. 60 p. 77
G_1	finite continuous group of motions	p. 68
G^i, G^i_h	fns. leading to Cartan's and Berwald's connection parameters	Eq. (4.5.3), pp. 30-31
$\bar{G}^i$, $\bar{G}^i_j$	counterparts of above functions in (i) projectively deformed $\bar{F}_n$, also used in (ii) conformally deformed space $\bar{F}_n$	pp. 51-52, p. 62
G^{ij}, $\bar{G}^{ab}$	relative symmetric tensors in F_n and	pp. 63-64

	conformally deformed space $\overline{F}_n$	
G^i_{kh}	Berwald's connection parameters in F_n	p. 20
$\overline{G}^i_{jk}$	" " in (i) projectively deformed $\overline{F}_n$, also used in (ii) conformally deformed space $\overline{F}_n$, and in (iii) infinitesimally deformed $\overline{F}_n$	p. 52, p. 62, p. 91
G^i_{jkh}	directional derivative of G^i_{kh}	p. 20
$\overline{G}^i_{jik}$	directional derivative of $\overline{G}^i_{ji}$	p. 52
h_{ij}	angular metric tensor	p. 21
h^i_j	mixed tensor	Eq. (12.4.1)
H, $\overline{H}$	Berwald's scalar curvature in F_n, $\overline{F}_n$	Eqs. (4.9.20), (5.2.3)
H-CC	curvature collineation for Berwald's curvature tensor	p. 99
H_k	covariant vector derived from H^i_{ikh}	Eq. (4.9.20)
H^i_j	Berwald's deviation tensor in F_n	Eq. (4.9.16)
$\overline{H}^i_j$	" " in (i) projectively deformed $\overline{F}_n$, and (ii) conformally deformed space $\overline{F}_n$	p. 55, p. 63
H_{kh}	analogue of Ricci tensor (of Riemannian geometry) derived from H^i_{ikh}	Eq. (4.9.20)
H^i_{jk}	tensor leading to Berwald's curvature tensor	Eq. (4.9.14)
H^i_{jkh}	Berwald's curvature tensor	Eq. (4.9.13)
$\overline{H}^i_{khj}$	" " in (i) conformally deformed space $\overline{F}_n$, also used in (ii) infinitesimally deformed $\overline{F}_n$	Eq. (5.3.22), p. 91
HR - F_n	recurrent Finsler space	p. 119
HR^2 - F_n	bi-recurrent Finsler space	p. 160
HS^2 - F_n	bi-symmetric Finsler space	p. 105
J	a scalar function	p. 201
J^i_h	a mixed tensor	Eq. (8.3.7)

K	Gaussian (or second or specific or measure of) curvature of a surface	p. 6
K-CC	curvature collineation for Cartan's curvature tensor K^{i}_{jkh}	p. 100
K^{i}_{jkh}	curvature tensor leading to Cartan's 3rd curvature tensor R^{i}_{jkh}	Eqs. (4.7.10), (4.7.13)
$\overline{K}^{i}_{jkh}$	" " in conformally deformed space $\overline{F}_n$	p. 64
$\widetilde{K}^{i}_{kh\,m}$	Rund's relative curvature tensor (super flex tilde sign used here not standing with $\widetilde{\Omega}(\overline{x}^k, \dot{\overline{x}}^k)$ used on p. 77)	Eq. (4.7.2)
l_i	unit vector along element of support $\dot{x}^i$	p. 20
L	a scalar function	Eq. (8.3.10)
M_k, M^{i}_{j}, M_{kh}, M^{i}_{jk}	contracted parts of tensor M^{i}_{jkh}	p. 188
N^{i}_{jkh}	normal projective curvature tensor	p. 55
N_{kh}	contracted part of N^{i}_{jkh}	p. 184
NP-F_n	normal projective Finsler space	p. 162
p_i	a covariant vector-field	p. 147
$p^{a}(x)$	fields of π-messons	p. 220
$((p^{j}_{i}))$, $((q^{j}_{i}))$	square matrices formed by the coefficients	p. 211
P	scalar function	p. 176
P-CC	curvature collineation for projective curvature tensor	p. 100
PHR-F_n	HR-F_n admitting a projective motion	p. 148
P_j, P_{jk}	covariant vector and tensor-fields	pp. 22, 176, 177
P^{i}_{jk}	a mixed tensor (in contrast to connection parameters P^{i}_{jk} defined by Eq. (4.4.1))	p. 22
P^{*i}_{jk}	Rund's connection parameters	p. 29
P_{jkh}	(v) hv-torsion tensor	p. 22

P^{r}_{ikh}	Cartan's second curvature tensor	p. 22
P_{ijkh}	Cartan's second (associate) curvature tensor	p. 22
PS-F_n	projectively symmetric Finsler space	p. 105
Q	scalar function	Eq. (11.4.30)
Q^{i}_{jkh}	curvature tensor like projective quantities (but not forming a tensor)	p. 54
$\overline{Q}^{i}_{ikh}$	" in deformed space $\overline{F}_n$ under an infinitesimal transformation	p. 85
Q_{kh}	contracted part of Q^{i}_{jkh} - an analogue of Ricci tensor	p. 96
$\mathbf{r}_{,1}$, $\mathbf{r}_{,2}$	tangent vectors to parametric curves	p. 14
R	some region in space	p. 25
R	(scalar) Riemannian curvature of F_n	Eq. (4.11.1)
R-CC	curvature collineation for Cartan's third curvature tensor	p. 100
R^{i}_{jkh}	components of curvature tensor, also used for Cartan's 3rd curvature tensor	p. 17 p. 37
R_{jkhl}	Cartan's 3rd (associate) curvature tensor	p. 40
RNP-F_n	normal projective Finsler space of recurrent curvature	p. 162
s	arc-length of a curve	p. 28
S	set of linearly independent vectors of vector space	p. 211
S-F_n	symmetric Finsler space	p. 105
SHR-F_n	special *HR*-F_n	p. 119
S^{i}_{jkh}	Cartan's first curvature tensor	p. 22
$\overline{S}^{i}_{jkh}$	" " in conformally deformed space $\overline{F}_n$	p. 64
S_{kh}	analogue of Ricci tensor (of Riemannian geometry) derived from S^{i}_{jkh}	Eq. (4.9.11)
S_{ijkh}	Cartan's first (associate) curvature tensor	p. 22

$SNP\text{-}F_n$	symmetric normal projective Finsler space	p. 162
t	a general parameter in the eq. of a curve	p. 23
T	a scalar function	p. 206
T^{ij}, $\overline{T}^{ij}$	components of a vector **T** in V^2 w.r.t. bases $\{\mathbf{e}_{ij}\}$, $\{\overline{\mathbf{e}}_{ij}\}$	pp. 214, 215
$T_n(\mathrm{P})$	tangent manifold at the point P	p. 18
$T_{(jk)}$	symmetric part of tensor T_{jk}	p. 48
$T_{[jk]}$	skew-symmetric part of tensor T_{jk}	p. 48
$T^{i_1 i_2 \ldots i_r}_{j_1 j_2 \cdots j_s}$	a mixed tensor of type (r, s)	p. 80
(u^1, u^2), u^λ	parametric (curvilinear) coordinates of a point on a surface	p. 13
$u^\lambda = \text{const.}$	parametric curves	p. 14
U_j	a covariant vector field	p. 55
V, V^*	non-null set of vectors forming vector spaces	pp. 209, 213
$V^* \times W^*$	Cartesian product of dual (vector) spaces	p. 213
$V \otimes W$	tensor product of vector spaces V and W	p. 213
V^1	vector space	p. 210
V_1	dual (vector) space	p. 212
V^2	tensor product of vector space V^1 with itself, i.e. $V^1 \otimes V^1$	p. 214
V_2	(i) tensor product of dual space V_1 with itself, i.e. $V_1 \otimes V_1$; also used for (ii) 2-dimensional curved surface	p. 215 p. 13
V^1_1	tensor product of dual space V_1 and space V^1, i.e. $V_1 \otimes V^1$	p. 216
V^r_s	tensor product of space V^1 with itself repeated r times with its dual space V_1 repeated s times, i.e. $V^1 \otimes V^1 \otimes \ldots \otimes V^1 \otimes V_1 \otimes V_1 \otimes \ldots \otimes V_1$	p. 217
V_n	n-dimensional Riemannian space	p. 16
V_{kh}	second order covariant tensor	p. 184

W, W^*	non-null set of vectors forming vector spaces	p. 213
W^i_j	projective deviation tensor	p. 56
W^i_{jk}	projective tensor leading to W^i_{jkh}	p. 56
W^i_{jkh}	projective curvature tensor	p. 57
$WR - F_n$	projectively recurrent Finsler space	p. 123
$WR^2 - F_n$	bi-recurrent projective Finsler space	p. 163
x^i	coordinates of a point in V_n	p. 7
(x^i), $(\overline{x}^a)$	coordinate systems	p. 18
$\dot{x}^i = dx^i / dt$	components of tangent vector to curve C	p. 18
$(x^i, \dot{x}^i)$	line-element of a curve C	p. 18
$\ddot{x}^i, x''^i$	second order derivative of x^i w.r.t. t	p. 28
$x^{(p)i}$	p^{th} order (partial) derivative of x^i	p. 23
$X_{ij}, \overline{X}_{ij}$	components of vector **X** in V_2 w.r.t. bases $\{\mathbf{e}^{ij}\}$,$\{\overline{\mathbf{e}}^{ij}\}$	p. 216
y_i	a covariant vector-field	p. 26
(z^i)	complex coordinate system	p. 24
$(\overline{z}^i)$	conjugate complex coordinate system	p. 24

1.2. Greek symbols

Symbol	Used for	Reference
$\alpha_i, \overline{\alpha}_i$	components of vector $\boldsymbol{\alpha}$ w.r.t. bases $\{\mathbf{e}^i\}$, $\{\overline{\mathbf{e}}^i\}$	Eq. (14.3.6)
$\delta_{\lambda\mu}, \delta^\lambda_\nu, \delta^k_i$	Kronecker deltas	p. 16, 26
ε	infinitesimal constant means 'belongs to'	Eq. (5.4.1), p. 214
η	denoting infinitesimal change $x^i \to \overline{x}^i$	Eq. (5.4.1)
η^k	a contravariant vector-field	Eq. (8.7.6)
ϕ	a scalar function	p. 45
φ^β	four-component space-time spinor	p. 220

Φ	a relative scalar function	p. 63
γ_{ikj}, γ^{i}_{jk}	generalized Christoffel symbols of I and II kinds in F_n	pp. 27, 28
$\bar{\gamma}_{ikj}$, $\bar{\gamma}^{h}_{ij}$	" " in conformally deformed space $\bar{F}_n$	p. 61
Γ^{i}_{jk}	Cartan's non-symmetric functions leading to his connection parameters Γ^{*i}_{jk}	Eqs. (4.5.2), (4.5.4)
$(\boldsymbol{\lambda}^*, \boldsymbol{\alpha}^*)$	element of Cartesian product space $V^* \times W^*$	p. 213
λ_l	non-null covariant vector field	Eq. (8.1.1)
λ_{jhk}, λ^{i}_{jk}	Generalized Christoffel symbols for relative functions G^{ij}	Eq. (5.3.27)
Λ^{i}_{jk}	relative connection parameters	Eq. (5.3.28)
μ_j	a covariant vector-field	Eq. (5.9.1f)
μ_{jk}	A covariant tensor-field	Eq. (5.9.6)
ν_j	covariant vector-field	p. 180
ω	angle between parametric curves	p. 15
$\tilde{\Omega}(\bar{x}^k, \dot{\bar{x}}^k)$	'displaced' value of object $\Omega\ (x^i, \dot{x}^i)$ under the coordinate change $x^i \to \bar{x}^i$	p. 77
Ω^{i}_{jkh}	used for components of a curvature tensor	p. 43
$\Omega^{i}_{(jkh)}$	symmetric part of the geometric object Ω^{i}_{jkh}	p. 50
$\Omega^{i}_{[jkh]}$	skew-symmetric part of the object Ω^{i}_{jkh}	p. 50
Π^{i}_{j}	transvected part of Π^{i}_{jk}	p. 53
Π^{i}_{jk}	projective connection parameters in F_n	pp. 52, 53
$\bar{\Pi}^{i}_{jk}$	" " in (i) projectively deformed $\bar{F}_n$, also used in (ii) conformally deformed space $\bar{F}_n$, and (iii) in infinitesimally deformed $\bar{F}_n$	p. 53, p. 64, p. 85
Π^{i}_{jkh}	directional derivative of Π^{i}_{jk}	Eq. (5.1.15)
$\overset{(n)}{\Pi}{}^{i}_{jk}$	normal projective connection parameters	p. 161

ρ_j	a covariant vector field	Eq. (5.9.4)
σ_k, σ^h	gradient vector-field and its associate vector	p. 61
ψ	a scalar function of line elements	p. 60
ξ_k	a covariant vector-field	p. 154
χ	a scalar coefficient	p. 174

1.3. Miscellaneous symbols

Symbol	Used for	Reference
§	used for Section	p. 1
$\mathfrak{B}$	Berwald's covariant differential operator	Eq. (4.10.5)
$d^m\,\Omega$	variation in object Ω under coordinate change	p. 77
$d^v\,\Omega$	" " under infinitesimal transformation	p. 77
$\delta X^i / \delta t$	δ-derivative of vector field $X^i(t)$ w.r.t. t	Eq. (4.4.6)
$\partial_j, \bar{\partial}_j$	partial derivation operators w.r.t. x^j, $\bar{x}^j$	p. 53
Δ	area of a triangle	p. 6
∇	covariant differential operator	Eq. (4.6.4)
∇_i, ∇^i, ∇_X	Cartan's covariant differential operators (also for covariant derivation in Riemannian space)	pp. 17, 37, 12
$\nabla_k X^i$, $\dot{\nabla}_k X^i$	Cartan's covariant derivatives of the vector field X^i	Eqs. (4.5.14), (4.5.13)
$\Rightarrow$	implies	Eq. (3.14.2)
$£\,\Omega$	Lie derivative of object Ω w.r.t. infinitesimal transformation	p. 77
N	set of natural numbers	p. 13
$P_k X^i$	projective covariant derivative of X^i	p. 54
R_n	n-dimensional real vector space	p. 18
$\mathfrak{R}_k X^i$	partial δ-derivative of $X^i(x^j, \xi^j)$	Eq. (4.4.7)

$X^i_{;k}$	original notation for Rund's covariant derivative	p. 48
$X^i\|_k$, $X^i_{\|k}$	original notations used by Rund for Cartan's covariant derivatives	p. 49
$X^i_{(k)}$	original notations used by Rund for Berwald's covariant derivative	p. 49
$X^i_{((k))}$	original notation for projective covariant derivative of vector field X^i	Eq. (5.1.20)

§ 2. Abbreviations

Abbreviation	Full form	Reference
B.C.	before Christ	p. 1
CA-collineation	curvature collineation of special concircular form	p. 166
CA-motion	contra affine motion (generated by special concircular vector field)	p. 118
CC , *CC*s	curvature collineation(s)	p. 99
cf., Cf.	confer	p. 28
c.i.t.	concircular infinitesimal transformations	p. 74
const.	constant	p. 14
cycl.	cyclic interchange	p. 44
Eq(s).	equation(s)	pp. 13, 26
ibid	in the same source mentioned above	p. 230
op. cit.	as mentioned above	p. 49
p., pp.	page, pages	p. 84
resp.	respectively	p. 28
RHS	right hand side	p. 81
Theo.	theorem	p. 94
w.r.t.	with respect to	p. 42

INDEX

D

N

O

P

R

www.ingramcontent.com/pod-product-compliance
Lightning Source LLC
Chambersburg PA
CBHW071334210326
41597CB00015B/1452
9781925823806